PLANE and SOLID GEOMETRY ESSENTIALS

With
Everyday Decision Making

PLANE and SOLID GEOMETRY ESSENTIALS

JAMES ELANDER

To order additional copies of this book, contact:
Xlibris
844-714-8691
www.Xlibris.com
Orders@Xlibris.com
802243

PLANE and SOLID GEOMETRY
ESSENTIALS
With Everyday Decision Making

(Comments for students, teachers and parents.)

Rodin's THE THINKER
Legion of Honor Fine Arts Museum, San Francisco, CA
From ELANDER'S File

This text is written to provide the student with a background for the essentials of Plane and Solid Geometry, which was begun in grade school, and includes many applications in the various professions and trades, also preparation for the SAT, ACT tests and more importantly improving their life time valid Decision- Making Skills. This was the original reason for teaching geometry. This book is also written for math teachers and/or future math teachers. Research has indicated that Geometry and CRITICAL THINKING SKILLS have scored the lowest on college tests. The Decision-Making Skills as stated was the original objective for teaching geometry created by Socrates, Plato and many others in order to have a well functioning democracy. In other words, a valid Decision-Making informed voting citizen is a requirement for a democracy!

The following statement was posted over the entryway to Plato's Academy. (400 BCE).

LET NO MAN IGNORANT OF GEOMETRY ENTER HERE

Plato's statement reflects how important he felt Geometry was for entry into his adult critical thinking academy, which has influenced the decision as to why schools required students to study Geometry. (Geometry taught correctly will improve one's Decision Making Skills.) Plato knew this and Dr. H. Fawcett justified it in the 13th yearbook (1938) of the NCTM. This program also incorporates Dr. G. Polya's discovery methods to make the course more meaningful. The success of both methods depends on the teacher, but this text will aid the teacher by incorporating the two approaches. These meaningful methods provide for a teacher's creativity. (The SSMA yearbook *A Half Century of TEACHING SCIENCE and MATHEMATICS* is very informative. You may have to contact the organization SCHOOL SCIENCE AND MATHEMATICS, since the yearbook is out of print. Many university libraries have it.) These ideas suggested in the 1920s have been overlooked, unfortunately! One major suggestion was to treat Congruency as a subset of Similarity. This turns out to be a great time saver and is really a better approach. The author was able to cover Plane and Solid Geometry plus Conic Sections and Math Induction plus related Advanced Algebra topics in one year by using this innovation. This provided more time for other courses such as Calculus, Probability, College Algebra and some computer programing. Geometry and Logic and all valid decisions are based on undefined terms, defined terms, postulates, plus the interpretation of inverses, converses, and contrapositives. All should understand direct and indirect justification methods, plus understanding the use and misuse of basic Statistics and Poll results.

This book is the result of listening to colleagues, professional speakers, and reading Math history books, plus hundreds of students asking questions, making comments, and completing their homework. There is a unique Critical Thinking Test by Dr. R. Ennis listed in the bibliography. I have given this test to several hundred students, both high school and college. The students with averages of C or below improved the most on the Critical Thinking post test. Many students of all ages pursue the study of Geometry and/or Mathematics with indifference, but the method of teaching plus related bits of Math History can result in creating a lasting interest in Mathematics. **Try this** method, both parents and teachers will be impressed in aiding the students to not only better understandings, but to be filled with the "joy of learning". The students will never forget their teachers and will also appreciate their parents for their help in learning the essentials of Geometry!

Respectfully, *Jim Elander*
*E*mail:elanderje@gmail.com
Advisor: Olaf Elander

ADDITIONAL BOOKS BY THE AUTHOR:

TGIF MATH

A book containing over a 100 activities and problems to help teachers convert those hectic days, like before **Homecoming**, into ones to look forward to. Parents and other adults also enjoy these Thinking Activities. Basic Decision-Making Activities and weaknesses are included.

Basic High School Math Review

A book for students to review and prepare their Math material before taking the next course or entrance exam to college with the why approach instead of just a show how. It includes basic Decision-Making Skills also!

TEAM WORKING
For Better Math Students

A collaborative Approach for Parents and Teachers of students in grades 1-12 (Yes, it covers the mathematics in all 12 grades!) with interesting bits of Math History as to the why and how we learn Math, plus everyday Decision-Making Skills!

This is a new book for Parents and Math Teachers, which uses a collaborative team **approach** for helping students to understand and enjoy mathematics. It contains explanation as to **Why** and **How** plus bits of historical background, which makes the math content even more interesting. This book covers the Math in grades 1 to 12. Yes, all 12 grades! It is written so parents and teachers will be able to make it more interesting for the student. It is especially useful for Home School programs.

These books are available at any book store or directly from the publisher at 1-888-795-4274 ext. 7879.

For more information on any of the above:

Website:http://sites.google.com/site/mathfordecisionmaking

or

contact the author at: elanderje@gmail.com, Subject: BOOKS

The author's first Geometry text was published in 1992 and this new text is much improved and covers Plane and Solid Geometry Essentials plus Decision Making basics.

CONTENTS

Chapter 3

Chapter 4

CHAPTER 8

PLANE and SOLID GEOMETRY ESSENTIALS
With Everyday Decision Making
The Objectives
(For Parents, Teachers and Students)

Objectives: The first objective is to help your students learn and understand the **essentials** of Plane and Solid Geometry. Naturally, to better prepare students to pass the mathematics questions on college entrance exams and other types of thinking tests is an objective. A major objective is to prepare the student with the understanding of the Geometric applications for the trade skills used in everyday applications.

Valid THINKERS are made, not born, and the scores on entrance exams will improve by reviewing the information you may have forgotten! There is an old saying that has a meaning for students: ***Mathematics is not a spectator sport!*** What does that mean?

An important part of becoming a CRITICAL THINKER and a valid Decision Maker, besides asking questions, is to recognize words that need defining, and that all decisions are based on **definitions, assumptions, and previous accepted rules or laws (theorems).** (This will have more meaning as you cover the material.)

All persons are eager to be valid THINKERS, but that ability is not a gift, but is learned, and unfortunately forgotten if not practiced. Some sessions in the chapters will review the types of decision-making skills and methods using the content from geometry and problems that are similar to the ones used or misused in everyday situations. The weaknesses of Critical Thinking will be corrected by studying the weaknesses integrated with this text in the sessions.

This integrated content also **qualifies for a high school course in Plane and Solid Geometry,** or as a supplement to a Math Education course for future teachers. It is also a valuable addition for reviewing sessions in the Jr. or Sr. years in preparation for college entrance exams. The ways and means for helping students of all ages to become critical thinkers and improve their test scores will involve the following types of activities for arriving at conclusions.

The role of definition
Inductive reasoning conclusions from:

Trends
Polls
Observations
Illusions
Statistics
Use and Misuses
Ads (Interpretations)
Deductive reasoning conclusions from:
Direct and Indirect reasoning
Logic and Implications

Plato had posted at the entrance to his Academy in Athens (about 500 BCE) the following: (Plato was training his students to be informed citizens in order to participate in their first democracy.)

LET NO MAN IGNORANT OF GEOMETRY ENTER HERE
Eric T. Bell: *MEN OF MATHEMATICS*

Who was Plato? When did Plato live? Why did Plato and Socrates teach "geometry"? Why was Socrates put on trial and found guilty? Athens created the first democracy and realized a democracy must have an informed and valid Decision-Making set of voters to succeed. (To qualify as a voter at that time a person must own property.) The student should discuss valid Decision-Making skills with their parents and teacher again at the end of the course. Plato's statement has now been changed to: LET NO STUDENT EXIT HIGH SCHOOL IGNORANT OF GEOMETRY.

Geometry is still required by colleges and the trades schools. Plato's reason will be understood as a result of these sessions. The students should be encouraged to ask their questions, understand the answers (even with parents), write their summaries or reviews, and they will become better Decision Makers.

There is also a bibliography of selected books for students, teachers and parents.

Many of these books are in the school's or the public library.

LEARNING TAKES WORK and in subject matter material like in a sport, there is required practice called homework. The author has classified these as Activities.

The types of Activities are:

Class: (Suggested many activities be started in class.)

Discovery: (A general conclusion type of problem leading to a conclusion followed by justification.)

Homework: (Completed outside of class)

Challenges:

The Activities are identified by three digits, example 2.1.2 which means Chapter 2, Session 1, Activity 2. The teacher should make the final decision as to how the Activities are used to best fit their teaching methods and how to assign the problems. The answers to the problems in many cases are given. This provides quick feed back to the student as to right or wrong. (Some answers are intentionally wrong.) It should be pointed out to the student that the real understanding will be revealed on tests where the student will not have the answers. (Students: Defend your answers!) Each student should have a ruler, graph paper, science calculator, protractor and a note book or another means for keeping notes.

HERE'S TO THE JOY OF LEARNING.
Jim Elander, Author

(Former high school and college teacher, plus an author, now retired)

Email: elanderje@gmail.com

He is thinking!

PLANE and SOLID
GEOMETRY ESSENTIALS
Pre-Course set of Questions

Note: This probably does not sound inviting to the student to have a set of questions at first, when beginning a new program. The reason (Teacher should explain to the class.) is to provide the teacher and the student with a basis to obtain a measure of their learning progress. I suggest the teacher make copies of the Pre-course Review and use class time to let the students complete the Review, Then grade it (in class) and collect their review and keep the copies so the student can compare their score with the post review test score at the end of the course to see their progress.

Teacher: It is suggested you collect the Pre-Course sheets and inform the class the sheets will be returned at the end of the course. There are 21 questions. The Parents, Teacher, and the Students will not be disappointed with the Post-Course results when compared to the Pre-Course scores.

The review is on the next three pages. This is so the teacher may easily duplicate the review, hand it out, students work it and, score it, collect it and then return it at the end of the course.

PRE-COURSE REVIEW QUESTIONS
(may use calculator)

1. The map of a triangular lot has the measurements 75 feet by 125 feet by X feet. The X measurement has been blurred, but you know X is greater than and less than how many feet. What are your answers?

2. Complete the following blank row with the logical set of numbers.

1

1 1

1 2 1

1 3 3 1

1 4 6 4 1

— — — — — —

3. What is the sum of the interior angles in this figure? (No protractor.)

4. Why does a 4-legged chair sometimes wobble?

5. If 2 points will determine one line segment, then how many segments could be drawn using the following? (In all cases no three points are in a straight line.)

   a. 3 points   b. 4 points   c. 5 points   d. n points

6. How many words in the following statement are undefinable? List the words that are undefinable. One such word is **the.**

   Statement: **Now is the time for all good men to come to the aid of their country.**

7. Is the following a valid definition? A dog is a 4-legged animal.

8. In the following right triangle, what is the measure of AD?

   Given: AC = 5 units CD = 3 units

9. All congruent triangles are similar, but all similar triangles are not congruent. True or false?

10. Are these two triangles similar?

    Triangle A has sides of 4, 7 and 5.
    Triangle B has sides of 10, 8 and 14.

11. The gate in the following figure is sagging. How would you fix it and what would you add?

12. If each letter represents a unique digit, then what digit is B equal to?

$$
\begin{array}{r}
\text{H I T} \\
+ \ \text{H I T} \\
\hline
\text{B A L L}
\end{array}
$$

13. In the following figure, how would you know if the pole is perpendicular to the ground?

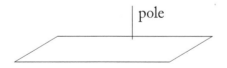

pole

14. What is a theorem?

15. What is a postulate?

16. In the following figure: A circle inside a square and the side of the square is 10 inches

   a. What is the perimeter of the square?
   b. What is the area of the square?
   c. What is the circumference of the circle?
   d. What is the area of the circle?

17. What is the volume of each of the following figures?

   a.  L = 10 units   W = 4 units   and H = 6 units

   b. Radius = 5 inches
      Height = 10 inches

   c. A ball with a 4-inch diameter.

18. If we assume the following statement is valid, then which of the following statements (a, b or c) do you think must be valid?

   Statement: If students do their homework, then they will pass.

   a. If students pass, then they did their homework.
   b. If students do not do their homework, then they will not pass.
   c. If students do not pass, then they did not do their homework.

19. What are the values for the mean, mode, and the median for the following set of numbers?   (5, 8, 9, 9, 12)

20. Which of these two classes would you say is better from these scores?

   Assume an A grade is 4 points, B is 3, C is 2 and D is 1.
   Class AM: Class PM: Your answer. _____
      A's   5   3
      B's   4   6
      C's   1   2
      D's   2   1

Now complete the following:

a. Draw a graph, using the scores, for each class.
b. The superintendent said the two classes were equal since they have the same average. What is the mean or the average for each class?
c. What is the median score for each class?
d. What is the mode score for each class?
e. Which class do you think is the better? Why?

21. Given the right triangle ABC with the following:

Angle A is 30 degrees and side AB measures 6 and angle ADC is 90 degrees, Then:

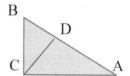

a. Angle B's measure is?
b. Angle DCB measure is?
c. CA's measure is?
d. DA's measure is?

Student's name: _____

Score: a. Number of correct answers. _____
          b. Number of wrong answers. _____
          c. Number of total answers. _____

Percent right is equal to X% where X equals (n/39)(100) rounded to an integer. Where n is the number of the students correct answers. 39 possible answers (teacher's total number may be different).

Answers to Pre-Course Review

1. $50 < x < 200$
2. 1   5   10   10   5   1
3. 360 degrees
4. 4 planes
5. a. 3   b. 6   c. 10   d. $n(n-1)/2$
6. The number of Undefined is between 5 and 9. The two that probably need defining are "good" and "aid."
7. No
8. 4 units
9. True
10. Yes
11. Lift the gate to form a rectangle and then add a diagonal.
12. One (1)
13. Check for perpendicularity from two directions.
14. A proven important mathematical statement.
15. An Assumption
16. a. P is 40 in.
    b. Area is 100 sq. in.
    c. circumference (circle) is 10(3.14) = 31.4 in.
    d. Area (circle) is 25 (3.14) = 78.5 sq. in.
17. a. Vol. of 240 cu. units
    b. Vol. of cylinder is 250(3.14) or 785 cu. inches
    c. Vol. of ball is 33.49 cu. Inches or 32(3.14)/3
18. "c" is valid.
19. Mode is 9. Mean is 8.6. Median is 9.
20. a. graphs (Teacher should check the graphs.)
    b. Mean: am. class is 3 or B. pm. class is 2.9 or B.
    c. Mode: am. class is A. pm. class is B.
    d. Median: am. class B. pm. class B.
    e. Which is the better class?

21.

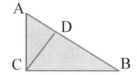

a. Angle B's measure is 60 degrees?
b. Angle DCB measure is 30 degrees?
c. AC's measure is 3. d. DB's measure is1.5.

PLANE AND SOLID GEOMETRY ESSENTIALS

Chapter 1
Session 1
Basics

MATHEMATICS IS THE GATE AND THE KEY TO ALL SCIENCES. HE WHO IS IGNORANT OF IT CANNOT KNOW THE THINGS OF THIS WORLD.

Roger Bacon

All college entrance exams contain questions related to Plane and Solid Geometry, but many students have forgotten much of their Geometry and Algebra. So, this course, will contain geometry questions similar to those on entrance exams and also cover the essentials in **Plane** and **Solid** Geometry, plus practical applications and topics related to Decision Making or Critical Thinking in everyday situations. (Socrates and others several hundred years B.C.E. used basic geometry to teach their students the basics of critical thinking or decisions making.)

In the Pre-Course Review, you were given a few examples of the mathematical problems and a **reminder that all everyday conclusions are based on undefined terms, defined terms, basic assumptions and previous conclusions (deductive reasoning), plus drawing conclusions from a few previous cases (inductive reasoning). (Do you understand undefined terms?) The above will be explained in detail and will be better understood as the course progresses.**

Example:

U.S. Constitutional Amendment XIX.

The right of citizens of the United States to vote shall not be denied or abridged by the United States or by any State on account of their sex. The following questions refer to the highlighted statement above.

(The author's answers are included. Your answers may differ some.)

What does "abridged" mean? (This may be an undefined term to the student.)
What is the Total Number of words? # _____ 29
How many words like **in** or **the** would you classify as Undefined?# _____ 16
How many words would **you** classify as Defined?# _____
What is the percent of the total number of words you classified as undefined?

Hint: The number of "undefined' words is what percent of the total number of words? (examples of undefined: the, of, to, etc.) Write the question in equation form and solve. Your number of Undefined words is x% of total of total number of words?

$$16 = x\%(29) \text{ or } 16 = x(1/100)(29)$$
Where did the 1/100 come from?
Solving: x = ?

Two general questions.

Questions: 1. What number (numeral) does the symbol % present?(1/100)
2. What number (numeral) does XIX represent? 109 or 19

Many conclusions are based on what a person sees. The following picture will illustrate this. Look at it from the right and then from the left. Which do you see, the face of an old woman, or a young woman? What if you were a witness to an accident, could you claim you saw something different from what another person saw?

W. Hill
Puck magazine 1915

What is seen may be reported differently by two witnesses!

Definition 1: A logical system is based on undefined terms, defined terms, assumptions or postulates, and theorems or conclusions which are justified by the previous four, and of course other decisions or laws resulting from them.

The word Theorem in the above definition can be defined.

Definition 2: A theorem is an important mathematical statement which can be proved.

Definition 3: A definition is valid, if it is valid when reversed.

You will understand the meaning of the above definitions by the end of this course. We will develop a logical system using simple geometric elementary concepts to help you understand and appreciate such a simple system. **Conclusions are based on Undefined and Defined terms, Postulates, and previous Conclusions by direct and indirect methods. An important conclusion in Geometry is called a Theorem.**

Note: As you may expect, elementary Geometric systems have become very complex over the years. Elementary Geometry is taught and required for two reasons:

1. It is used in many professions such as construction, navigation, medical, trades, etc.
2. The logical thinking or decision making (conclusions) training is required by basically all people, especially in a democracy.

It was recognized about 2000 years ago that geometry was an easy way to teach youth how to make logical decisions. (Socrates and others in the city of Athens which resulted in Euclid writing the first geometry text.) Geometry today is basically taught to all students for the above two reasons.

Following are 6 postulates for this Geometric system. (more will be added when needed.) What is a postulate?

Postulate 1: A line has an infinite set of points.

Postulate 2: Two points will determine one and only one straight line.

Postulate 3: There is a one to one correspondence between the points on a line and the real number line.

Postulate 4: The shortest distance between two points is the measure of the straight line segment determined by the 2 points.

(This is not always true as every taxicab driver knows.)

Example: A taxicab driver has two options to travel from A to B.)

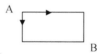

Postulate 5: The shortest distance between a point and a line is the perpendicular line segment.

What is a Geometric Plane? Your teacher will answer the question and no doubt point out a few examples of planes in the classroom.

Postulate 6: Three non-collinear points determine one geometric plane.

(List some examples of plane segments items in your house.)

PLANE AND SOLID GEOMETRY ESSENTIALS

Class Activity 1.1.1 (Chapter 1, Session 1, Activity 1)

1. Use your ruler to draw the following triangles, with sides a, b, and c as given. (centimeters.) Parent: You may teach your student an easy way using a (math) compass as you may have done in school.

	a	b	c
a.	3	3	5
b.	4	4	4
c.	3	4	5
d.	6	3	4
e.	6	1	5

2. From the measurements in number "1" complete the following:

   a. The sum of the measures of any two sides (segments) of a triangle is greater than ...
   b. Which postulate is this based on?
   c. Try to draw the triangle with sides of 2 in, 1 in, and 3 in.
   d. Try to draw a triangle with sides of 2 in, 4 in, and 1 in.
   e. If a triangle has sides 3,5 and N, then what do you know about side N? Hint: N must be great than what value and less than what value?
   f. If a + b > c, or b +c >a or a + c > b, are these all valid for any triangle?
   g. Using the results from number 2a, and given a triangle with sides 10, 18 and a, then what do you know about the length of side a?
   h. Write the inequalities in #2g using the term "sides" of a triangle instead of a, b, and c. Label this:

Theorem 1. The sum of the measures of 2 sides of a triangle is greater than the measure of the third side.

Hint: The justification follows from postulate 5 therefore, we know that
$$a + b > c, a + c > b, b + c > a.$$

Can you use algebra from $a + b > c$ to arrive at $a > |c-b|$. Write the 3 forms similar to 1g using the absolute value symbol.

Problem: (Condition: The cab must be always moving in the direction of B.)

3. In the following 2 cases how many ways can the taxi go from A to B?

    (Condition: The cab must be always moving in the direction of B.)

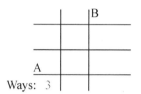 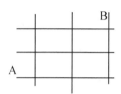

Ways: 3   Ways: 6

4. Complete the following and then write a conclusion. (use a calculator)

    a. 9 times 1917=? Add the digits in the product until you have a one-digit number.) 17253 to 18 to 9

    Answer: The product in 'a' is 17253. Adding results is 18 then 9.

    b. 9 x 2468 = ? Adding results = ?
    c. 9 x 13579 = ? Adding results = ?
    d. 9 x 968573 = ? Adding results = ?
    e. Write your assumed conclusion? (A number multiplied by nine always adds to?
    f. From your results, do you think that 9870654321 is divisible by 9?

    In this session the examples illustrated that conclusions are arrived at by using inductive reasoning. Write what you think a definition of Inductive Reasoning is:

5. Give an example or a case where the perpendicular distance is the shortest distance, and where it is not the shortest.

6. [(5x+7y) – 10y]z = 225, then which of the numbers x, y or z cannot be zero?

Answers: Activity 1.1.1.

1. Cases a, b, c, d, are possible but e isn't.
   f. a + b > c, a + c > b, b + c > a   g. yes   h. 8 < x < 28

3. 3 ways, 6 ways
4. f. Yes, since the sum of the digits is 45 which add to 9. This process is called Casting out nines! Inductive reasoning!
5. Point and line on a geometric plane. Commercial airlines path to a distant points like Chicago to Paris.
6. Z cannot be 0

Another example below illustrates that what you see may be different from what another person may see.

From the above, can you tell the difference between deductive and inductive reasoning?

Two examples: Problems number 3 is deductive and number 4 is inductive. What do you see in the picture below?

A friend from Norway sent this to the author.
From Elander's File

Record your summary.

Suggest the students keep a summary, which is a valuable learning method as to what they think is important to remember. Example: The definition of a Theorem?

Definition 2: A theorem is a mathematical statement that has been justified or proven.

What does a logical system consist of?

You will learn how conclusions are arrived at by the Direct and Indirect methods. (Inductive and deductive reasoning)

Teacher: Read the 13th yearbook of NCTM related to teaching Decision Making in Geometry. (You may have to check the closest college or University library for the book.)

Note to teacher and student!

In some of the activities the answers are wrong! This gives the students an opportunity to show they are correct!

PLANE AND SOLID GEOMETRY ESSENTIALS

Chapter 1
Session 2
Basics #2

YOUNG PEOPLE WHO HAVE ACQUIRED THE ABILITY TO ANALYZE PROBLEMS, GATHER INFORMATION, PUT THE PIECES TOGETHER TO FORM TENTATIVE SOLUTIONS WILL ALWAYS BE IN DEMAND.

J. G. Maisonrouge
Board Chairman
IBM World Trade Corp.

As stated in Session 1, college entrance exams all contain questions related to Geometry and Algebra, but many students have forgotten most of their Geometry and Algebra due to lack of use. Just like you have forgotten some math from earlier grades. So, we will start with a review of Geometry and Algebra and develop the problem solving methods, incorporating the types mentioned in the Objectives.

In Session 1 you were given a few examples including undefined terms, algebra, guessing, etc, plus introduced to the concept of drawing generalizations from a few cases. In this section a few more undefined terms and a few postulates will be provided. (Do you understand what a postulate is?)

A geometric plane will be classified as an undefined term, but you really know what a geometric plane is from past experiences. The following figure is a drawing of a plane segment.

Some items that consist of plane figures are: table tops, floors, sheets of cardboard, window panes, cover of a book, walls in your room, etc. Three-dimensional objects like boxes consist of plane figures. In the picture of a box below, how many plane segments are used to create a real box?

The answer is 6.

Hint: Draw the above figure flatten out, called the layout view, and you will see the six-plane segments. Better yet, take a box and actually flatten it out and look at the six-plane segments.

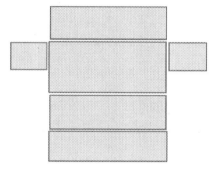

Discovery Activity 1.2.1

(Chapter 1. Session 2. Activity 1.)

1. Take a plane segment like a stiff piece of cardboard or a book.

   a. Can you balance or support it with a tip of one finger?
   b. Can you balance or support it with the tips of two fingers?
   c. Can you balance or support it with the tips of two fingers and your thumb? Yes
   d. Repeat a, b, and c with another plane segment. (Like a mirror, but don't drop it!) Write a general statement as to your conclusion. You may have observed the following postulate and why it is a postulate.

   Postulate 6: Three non-collinear points determine one geometric plane. Why the word non-collinear?

2. Draw the three-dimensional view of a pyramid with triangular base. How many plane segments are there? Draw a layout view of a pyramid. (Need a dictionary?)

3. The reverse problem is to take the flattened out or layout view and draw the three-dimensional view. Draw the 3-D view for the following. Name what you think each could represent.

4. Postulate 6 states that three non-collinear points determine a geometric plane. How many planes could four points determine?

   Hint: Label the points A, B, C, and D and list the planes, three points at a time.

5. Why is it possible for a 4-legged stool to wobble? Write out your explanation and explain it to your parents. (This is an application of problem 4.)

6. Two points will determine one line, then how many lines will three **non-collinear** points determine?

    a. How many lines will four non-collinear points determine?
    b. How many lines will five non-collinear points determine?
    c. How many lines will six non-collinear points determine?
    d. Do you see the pattern and predict how many lines will N non-collinear points determine?

Think Problem: Tom lives on Euclid St. in the small town of Educateville and walks to school. Going to school he passes a gas station, a U.S. Post Office, a video store, a fast food place, and the school's football field, but not in that order. Going home from school, he passes the Video store before the football field, but after the Post Office. He passes the fast food first. He also passes the gas station second. On the line below place in order the five places he passes.

Activity 1.2.1: Answers:

1. Fingers ---planes
2. 4 planes.
3. Layout to 3D figure
4. ABC, ABD, ACD, BCD,
5. Application of #4
6. Complete the following table and look for a pattern.) There is a pattern.

2 points	1 line	1
3 points	3 lines	3
4 points	6 lines	4
5 points	10	5
6 points	15	?
10 points	?	

Conclusion: N(N-1)/2

Challenge Program

You have a garden in your back yard, it measures 30 by 30 feet with 5 posts on each side. Your father gives you the credit card and tells you to buy the posts at the lumber yard. How many posts will you buy?

Answer to Think Problem question:

Home   FF   VS   PO   GS   FFS   School

Answer to Challenge problem is 16 posts.

Write your summary
(Some Suggestions)

Record the postulates for quick reference and the page number each is on.

Notice: The answers will be given to many of the problems. This is an aid for learning and fast checking for errors, plus an aid to creating the questions WHY and HOW. Students like having the answers, but they must understand since the proof as to their understanding will be on tests where they won't have the answers!

PLANE AND SOLID GEOMETRY ESSENTIALS

Chapter 1
Session 3
Angle Triangle Sum

You have all seen railroad tracks and they resemble what we call parallel lines, but in Geometry lines are required to be straight. So, two parallel lines in geometry are illustrated below.

Parallel lines on a plane (Plane Geometry) will never intersect, no matter how far extended. This is not true on other surfaces. Example: Pick any two points on the globe at the equator and travel due north on the globe from each point and where will the two lines intersect?

Class Activity 1.3.1

Draw the similar figure on your paper using your ruler. Measure $\angle 1$ and $\angle 2$ using your protractor. What do you and the class observe relative to angles 1 and 2.

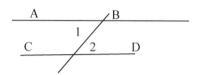

Write your conclusion in "If- then" form. Check your conclusion with those made by your classmates and teacher.

In Plane Geometry, the above conclusion is call an assumption or postulate, since we can't prove it, but it seems true by measurement. Label it as Postulate 7. (Students may suggest the measurement proves it, but nothing can be measured exactly.)

Postulate 7: If two parallel lines are crossed by another line (called a transversal), then the alternate interior angles are equal.

What is a postulate?

Class **Activity** 1.3.2: Angles

You probably learned how to measure angles in previous math classes, but as a reminder the method will be reviewed. The tool for measuring angles is a protractor. One type is illustrated below, but first some review (students probably had this in previous classes). **The following Activity is a review.** An angle consists of two rays with the same beginning point.

The method for labeling angles is using three letters, one letter on each ray and one at the beginning point. The beginning point is called the vertex.

Example: Angle ADG, where the middle letter is always at the vertex. Notice the symbol for angle is ∠. The drawing for ∆ADG is:

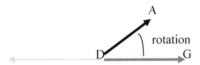

An angle is measured by the amount of rotation by using a Protractor.

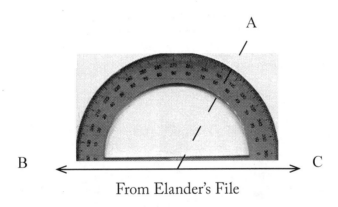

From Elander's File

Comment: Teacher may need to review how the protractor is used.

A bit of History: The ancient Babylonians assumed the complete circle rotation was 360 degrees from the approximate number of days in a year: Therefore, a half circle has 180 degrees. (The Chinese had the same idea.) An angle that rotates a half circle forms a straight line. The angle ∠COB forms a straight line. Angles are measured in degrees. A degree is $1/360^{th}$ of a circle. A degree is divided into 60 minutes (From the word mi' nute, meaning small parts.), hence the name minutes. A minute is divided into 60 seconds, the second set of minute parts, hence the name seconds.

Review of Angle Measure Summary

Teacher: It is assumed the students had these in previous grades.

An angle of 360 degrees (symbol for degree is °) is a complete revolution of a circle.

A straight angle is 180°.

A right angle is 90°.

1° equals 60 minutes (symbol for minutes is ')

The first set of minute parts.

1 minute equals 60 seconds (symbol for seconds is ")

The second set of minute parts.

An Acute angle is defined as one less than 90° and greater than 0°.

An Obtuse angle is defined as one less than 180° and greater than 90°.

An angle of 17° 10' 5" would be read as seventeen degrees, ten minutes and 5 seconds.

Teaching note: If the students need help using degree measurements review a few each day (operation +, -, x,) with a few problems for a week or two.

Now a method to arrive at some necessary conclusions: These will be a first in this review and will illustrate the value of Proof by Deduction or a Direct Proof.

We will start with two known figures, parallel lines and the simple figure called a triangle. What is your definition of the following from past education? You no doubt have had these terms previously. (Let the students develop the definitions, but the teacher should make sure the definitions are correct.)

Write your definition for: Ray, Segment

Definition 4: Parallel lines are lines in the same plane that do not intersect.

Definition 5a: A triangle is a set of 3 non-collinear points and the three line segments determined by the 3 points.

Definition 5b: An angle is 2 rays with the same beginning point.

(Remember the test for a valid definition is that it is true when reversed!)

Label your definitions of parallel lines and a triangle as Definitions 4 and 5.

Class Discovery Activity 1.3.3

Draw a similar figure to the one below on your paper. (Use your ruler.)

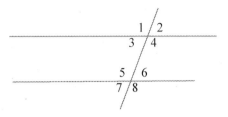

For convenience, the angles are labeled with numbers and from the **Alternate Interior Angle Postulate** 7, we know the m∠3 equals m∠6 and m∠5 equals m∠4.

(The "m", reads measure of the angle and is put in front of the angle symbol ∠, so that the statement is true, number-wise.)

Now more conclusions! In the figure above, if the measure of angle 3 plus angle 1 equal 180 degrees and angles 1 plus 2 equal 180 degrees, then what can you say about angles 3 and 2?

m∠3 + m∠1 = 180° Why?
m∠1 + m∠2 = 180° Why?
Hence: m∠3 = m∠2 By algebra

Opposite Angle Theorem

Theorem **2:** **If two lines intersect, then the measure of the opposite angles is equal and the angles are congruent, or identical. (See figure above.) m∠1 = m∠2 and therefore ∠ = ∠2 and conversely.**

Read again Definition 5 (pg. 17) which state what an important conclusion in Mathematics is. (What does the above conclusion mean?) This is why the above statement is labeled Theorem 2: notice the theorem is if-then form.

JAMES ELANDER

Class Discovery Activity 1. 3.4

1. In the center of your paper locate a point and label it P.

2. About two inches below the point P, using your ruler, draw a line and label it AB.

3. Now draw a line through point P that is parallel to line AB.

4. Can you draw another line through point P that is parallel to AB.

5. Did you succeed in # 4?

6. The ancient Geometers didn't succeed either, so they created the following postulate.

Parallel Postulate 8: Given a geometric plane with a line and a point not on the line, then through the point there is only one line on the plane that is parallel to the given line. (Euclid's parallel postulate.) The students may ask who Euclid is or was (perhaps a student would volunteer to give a report).

Home Activity 1.3.5

Given the following figure with two parallel lines intersected by a transversal. The angles are numbered. Draw a similar figure on your paper using a ruler

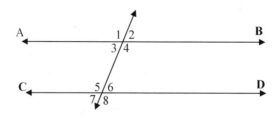

Give a reason for each of the following.
Answers

1. Why is m∠1 = m∠4?   Theorem 2
2. Why is m∠5 = m∠4?   Postulate 7
3. Why is m∠2 = m∠6?
4. Why is m∠3 = m∠7?
5. Why is ∠8 = m∠4?
6. Why is m∠5 = m∠1?
7. Why is m∠2 = m∠7?
8. Can you justify that m∠1 = m∠8?
9. If angle 1 is (3x - 20)° and m∠3 is (5x +40)°, then what is the measure of angle 6? Show your algebra!
   Answer: m∠6 = 80°
10. In the following figure it is given:
    Line AB is parallel to line CD, and PCD is a triangle. The numbers are the angles not the measure of the angles.

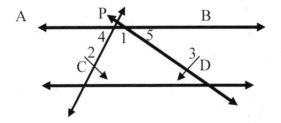

Draw the above figure, but larger, on your paper and measure angles 1, 2, and 3. What is their sum? Write your findings by completing the following: The sum of the angles in a plane triangle appears to be 180 degrees.

11. Can you prove that it is exactly 180 degrees. The following is the proof. Using the drawing or figure from #10, line AB is parallel to line CD and lines PC and PD are transversals and angles 4+1+5 equal 180 degrees. Why? Therefore: angles 1+ 2 + 3 equals 180°. Why?

Theorem 3: **The sum of the angles in a plane triangle is 180 degrees.**

Teacher: Ask if there are any questions. You may have to carefully go step by
step and help justify the theorems for the class.

How do the students define a triangle?
What do they think a polygon is? A quadrilateral, a square, a rectangle?

Class Activity 1.3.6

1. What is the sum of the angles in the following figures?

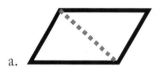

a.

Hint: In the figure above, what two figures are formed by drawing a
diagonal? Hence: The sum of the angles is ?

b.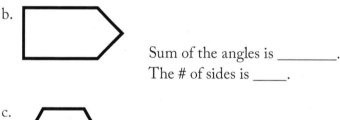

Sum of the angles is _____.
The # of sides is _____.

c.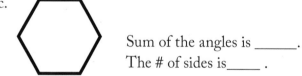

Sum of the angles is _____.
The # of sides is_____ .

2. Do you see a relationship between the number of sides and the number
times 180?

Hint: Complete the following for the figures above.
From the figures in problem 1. what is X in the following?

a. X times 180°   b. X times 180°   c. X times 180°

Conclusion: **Number of sides n ? (n-2) times 180° = sum of angles.**
Write your conclusion as a general statement. This a conclusion by Inductive Reasoning.
Teacher: You may need to explain Inductive Reasoning again.

3. In the following figure angle CBA is equal to how many degrees?

   Why? Hint: From Theorem 3 what is the measure of angle ABD?

   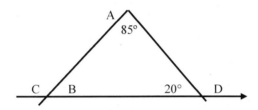

   (Angle CBA is called an exterior angle of a triangle. Write your conclusion as Theorem 4. (The theorem is also listed in the Summary of this session.) This result is by direct proof.

4. In the following figure It is given that angle AFC is equal to angle DFC. Is line AB parallel to line DF? Guess first, then try to justify your answer.

   Given:
   Lines AB, CG,
   with m∠AFC= m∠GCF

   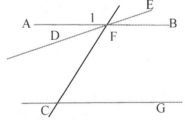

   Justify: What can you say about the DF line and AB?
   Teacher:
   Method: Use indirect proof (below) and justify that lines AB and FD are parallel, and if the lines are parallel then it must be the same line as AB.(Post. 8.) The line AB is the same as line DF by Post 8.

(Teacher: You will probably need to explain Indirect Proof which is a very useful method. The method of Indirect Proof is: All possible answers or cases are listed and all turn out false except one, then that one is the answer.

The following again will explain and illustrate the geometric case reasoning for the conclusion for the above problem #3. It is given original angles 1 and 2 are equal and also angle 2 equals angle AFC by Alternate Interior Angle Postulate 7. This is a contradiction of the postulate since there are two lines through point p that are parallel to CD (but there can be only one.), therefore AB and EF must be the same line or identical.

Teacher:	Ask for questions, if they have any. Most students do!
Theorem 5:	**If two lines are crossed by a third line so that the alternate interior angles are equal, then the 2 lines are parallel.**
Teacher:	There will be much more on converses!
	In geometry there are two types of proofs, Direct and Indirect.
Definition 6:	Direct proof method follows directly from the given, example:
	A = B = C, therefore A = C
Definition 7:	Indirect proof is the method where you list all the possibilities and show that all but one is impossible, therefore the remaining possibility is the correct one.
Thinking question:	A neighbor asks you to help build a fence around his back yard for a garden which will be a square 60 feet by 60 feet. He wants ten posts on each side of the square. You are instructed to take his truck and buy the posts at the lumber yard (Charge the cost). How many posts would you buy? **Think about it, even draw a picture of the problem to help you answer the problem.**
	Most students buy too many!
	This is a repeat problem to test memory.

Exam type question: Given: Angle A is 40 degrees, angle C is 20°, and angle ABD is 55 degrees. What is the measure of angle x?

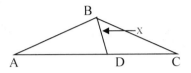

Answers to Activity 1.3.6:

1. a. 360°   b. 540°   c. 720°
2. (n-2)(180) where n is the number of sides.
3. Angle CBA equals 105°
4. Proof

Thinking Question: 36 posts

Summary
(List the postulates, definitions, theorems, and
other items you think are important.)

Definition 1: A logical system where the conclusions are based on undefined terms, defined terms and previous logical conclusions.
Definition 2: An important conclusion in mathematics that is proven or justified is given the name theorem.

    Students should add more to the summary with regard to what they think is important.

Example: Definition 7: Indirect proof is a method where you list all the possibilities and show that all but one is impossible, therefore the remaining possibility is the correct one.

In geometry you will study the figures called polygons which consist of a set of points and the line segments connecting the points. A triangle is an example.

Teacher: If your students don't recall the definition of a polygon help them create a valid definition.

Content in Chapter 1

There were 8 postulates which are...
There were 5 theorems which are...
There were 7 definitions which are...
What is a direct proof?
What is an Indirect Proof?
How to tell if a definition is valid??

Student: Suggestions, students may need a review of certain topics. (Write down some of the weaknesses.)

Write the items that you think are important as your summary for each chapter. This is a success habit of good students.

Angle summary (Review from former courses.)

An angle of 360 degrees (symbol for degree is °) is a complete revolution. A straight angle is 180°. A right angle is 90°.
1° equals 60 minutes (symbol for minutes is ′) Minute means small part. 1 minute equals 60 seconds (symbol for seconds is ″)
An angle of 17° 10′ 5″ would be read as seventeen degrees ten minutes and five seconds.

Challenge Problems
The Magic Square Problem

This problem was created by Chinese mathematicians about 300 BCE. The problem is to replace the ? marks with the a digit from the first 9 digits so that each row, column and diagonal adds to 15. (A digit can be used only once.) To make the problem easier the center number (5) is given for you.

?	?	?
?	5	?
?	?	?

Hint: Make a list of the possible sets of three digits that add to 15. Example: A diagonal could be (1 5 9) since 1+5+9 adds to 15. Since this is probably your first encounter with magic squares, then to make it a bit easier for you the middle digit is 5.

How old is Susan?

Miss M said today we are celebrating Susan's birthday. She will be the square root of 32 times the square root of 8 today. How old is Susan and what grade do you think she is in? (The answer is an integer.)

Room for extra problems the teacher may wish to add. (Challenge or review!)

Write your summary.

Thinking Problem

Replace the letters with integers from the following set, 0. 1-9, so the addition is correct.

Rule: an integer from the set can only represent one integer.
Example: If H is 4, then no other letter can be 4.

$$
\begin{array}{r}
\text{HIT} \\
+\ \ \text{HIT} \\
\hline
\text{RUNS}
\end{array}
$$

Question: What integer must R be? Is there more than one answer?

PLANE AND SOLID GEOMETRY ESSENTIALS

Chapter 2
Session 1
Similar Triangles

GOD gave us the integers (whole numbers) and all the rest is the work of Man.

L. Kronecker

Teacher: Dr. Geo. **Birkoff**. A Harvard mathematician (U. of Chicago Grad) was the first noted American Mathematicians, suggested that Congruency be taught as a subset of Similarity in order to save weeks of time. This is in the 5th Yearbook of the NCTM. I found this very useful in the 1950s to provide more time for subjects like A.P. Calculus.

The suggestion was never really adopted by textbook writers, except in only two cases, 1940 and 1992. As stated above, Congruency treated as a subset of Similarity is logical and does save time for more important material.

In this section you will work with figures which are subsets of polygons and learn some of their properties. A polygon is a figure consisting of points and the line segments determined by the points. A triangle is a polygon.

The definition of similar figures:

Definition 8: Two figures (polygons) are similar, if when mapped, the corresponding angles are equal, the corresponding sides parallel and the ratios of the corresponding sides are equal.

Remember: Definitions are valid when reversed! Write the reverse (called the converse) of the above definition.

An easy way to illustrate this Definition 8 is to map the polygons.

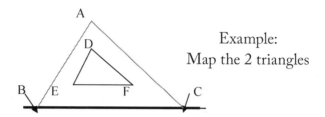

Example:
Map the 2 triangles

If the triangles are similar, then m∠A = m∠D, m∠B = m∠E, m∠C = m∠F, and AB/DE = 4/2 =2, BC/EF = 10/5=2, AC/DF = 12/6 =2. These two triangles are similar and the ratio of similitude is 2/1.

It is easier to justify similarity if the polygons are mapped such that the corresponding sides are parallel. In the following figure can it can be justified and explained why line EF is parallel to line BC? Question: Which Theorem states the condition for lines to be parallel?

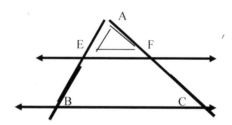

Question: If AB is 8 ft, AE is 2ft, and AF is 3 ft, and BC is 14 ft, then what are the measures of AC and EF to the nearest inch.

EF = 3ft. 6 in. or 3.5 ft. AC = 12

What word or words in the following need defining? Polygon

Definition 9: A polygon is a closed plane figure consisting of n points and the line segments determined by the n points, such that none of the line segments intersect, except at their end points.
Draw the diagram for a polygon consisting of 5 points!
Definition 10: A triangle is a three-sided polygon.

Teacher: Ask the students to recall the types of triangles they studied in previous classes. Write the definitions and draw the types of triangles. Discuss the following types of triangles, isosceles, equilateral, acute, obtuse. right, scalene. It can be assumed they had most of these types in previous classes and may just need a review.

With regard to triangles, if one of the following conditions exist, then the triangles are similar. (These conditions will be classified as postulates.) This is a worthwhile class activity with the teacher's help.)

Postulate 9: If two angles of one triangle are equal in measure to two angles of another triangle, then the triangles are similar. This is labeled the A A condition. (notice the ratio of corresponding sides is not known.) See Definition 8 on page 29.

Postulate 10: If in two triangles the corresponding sides are in equal ratio, then the triangles are similar and all the conditions in the general definition are valid. This is labeled the SSS condition. (Notice the measure of the angles is not known.)

Postulate 11: If two triangles have two pairs of corresponding sides in equal ratio and the **included** angles equal, then the triangles are similar and all the conditions in the general definition are valid. This is labeled the SAS condition.

The symbol that indicates that figures are similar is ~, like a lazy S.

Special case

Definition 11: Congruency: If two polygons are similar and the ratio of the corresponding sides is 1, then the figures are congruent.
What does congruent mean to you?

How would you describe congruent figures? Name a few congruent figures in your classroom or home. Mathematicians have a symbol to indicate this condition. The symbol for congruency is a combination of the equal sign and the symbol for similar figures. Congruency symbol is ≅.

Class Activity 2.1.1
(For Understanding)

1. Draw figures to illustrate each of the three conditions for triangle similarity. (Use your ruler and protractor.)

2. Given: ΔABC and ΔEDF (The order of the letters indicates the mapping.) with m∠B = 90° = m∠D

   AB = 6 and BC = 8   m∠C= 30° = m∠F
   DE = 3 and EF = 5

   a. Map the triangles.
   b. Are the triangles similar? Why?
   c. What is the measure of: AC? DF? and of angle A = angle D?
   d. What is the ratio of similitude?

   Answers: 2b. yes by Post. 9 2c. 10, 4, 60 degrees 2d. 2/1 or 1/2

3. Given: Triangles ABC and DEF with BC= 15, CA = 18, AB = 21, FE = 7, DE = 6, DF= 5

   a. Map and label the triangles so that the corresponding sides are in equal ratio.
   b. Are they similar? Why?

      Note: (Q.E.D. Quod Erat Demonstrandum) What does Q.E.D. mean? Students at one time always ended their justifications with Q.E.D.

   Answer: 3a. A-->E, B-->F, C-->D for mapping 3b. SSS ratio 1/3 or 3/1

4. Given: Triangles ABC and DEF with m∠F = 50° = m∠C, DF =10, BC = 5, CA = 7, and EF = 14.

   a. Map and label the triangles so that the corresponding sides are in equal ratio.
   b. Are they similar? Why?
   c. What is the ratio of the sides?

Answer: a. A --> E, B ---> D, C---> F b. Yes, by SAS c. 2/1 or ½

1. Given: In the following figure there are two parallel lines and the two intersecting lines forming two triangles with m∠AEF 50°. Lines AB and EC are perpendicular.

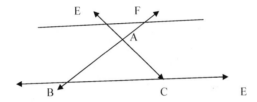

Hint: Map the two triangles. Are they similar? Why?
If they are similar, indicate the corresponding sides and Angles, map the triangles, and angle c is what value?.

Important problem: This problem is on most national tests. Right triangle ABC with C the right angle and angle A measures 30 degrees. (A right angle is one that has one 90 degree angle.)

Given: Angle CDA is also a right angle. Work in Class for discussion!

30 degrees A
90 degrees

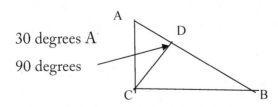

1. a. How many triangles are there? Name the 3 triangles using the three letters method.
   b. What are the degree measures of angles B, DCA, and DCB?
   c. Are there any similar triangles and if so, map them?

$$\Delta\ ACB \sim \Delta ADC \sim \Delta CDB$$

Note: The above examples used direct reasoning, meaning the conclusion followed directly, but indirect reasoning is actually used more than direct reasoning. Here is a unique application of indirect reasoning. You will not forget it, once you understand it! The teacher should explain again indirect reasoning as listing all possible cases and all but one is false, hence the remaining one is correct and valid. Give some examples.

The following is an excellent example of Indirect Reasoning, but may take time for all the students to understand!

A president of a corporation had three wise vice presidents, and naturally, the president wanted to know which one is the wisest, so he devised a test.

The Test: He placed one vice president (VP) in each corner of a large room and blind folded them. He told them that a black or white beanie would be placed on their head. Then the blindfolds will be removed and if they see a black beanie, they were to raise their right hand. When they knew what color their beanie is, they could take their hand down and state their answer, but if they are wrong they will be dismissed. (To prevent guessing they will have to explain their thinking.)

So, the president put a black beanie on each VP and told them to takes off their blindfold. They all raised their right hand. After a few minutes, one of the VPs took his hand down and declared he has a black beanie.

How did this VP know his beanie was black?

Hint: Use indirect reasoning. What are the possibilities?

Teacher comment: The easiest way for students to understand this **indirect reasoning** problem is to act it out in the class room. (Suggest the student use this as an after dinner family activity, and they can explain the solution to their parents.)

Solution hint:

Read the problem again and note the phrase "after a while." The VP who took his hand down knew his beanie was either black or white. Therefore, he assumed his beanie was white. If this were true the other two would know immediately that their beanies were black. Since they did not put their hands down, then his beanie had to be back! (Check the definition for Indirect Reasoning, Definition 7.

Activity 2.1.2
More practice?

1. The sides of a triangle are 12, 20 and X. Side X is between what two values? (side X is <? and >?)

2. A farmer wants to divide his pasture into two sections by drawing a line from side AB to side CD as shown by EF in the figure below

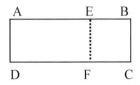

The measurements are AB is 720 yards and he wants the ratio of AE/EB to be 5/3. Hint: (720 -EB)/ EB = 5/3, therefore AE = ?

3. Draw an equilateral triangle with a side of 6 cm, and from each vertex draw the segments determined by each vertex and the midpoint of the opposite side. Label the vertices ABC and the midpoints DEF.

   How many triangles are there?   10

4. In the following figure it is given: AB is parallel to DC, AB is 15 units, DC is 9 units, and GF is 5 units. EGF is a straight segment and G and F are also midpoints. Complete the drawing on your paper. (Let the students explain their answer. There will probably be several answers.)

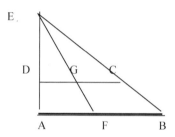

   Questions:
   Is triangle EDG similar EAF?

   a. If yes then map them and
   b. explain why.
   c. If your answer to "a" is yes, then write the ratios and use algebra to solve for the length of EF.

5. What is wrong with the following conclusion that 2 is equal to 1? Explain your answer!

   If X is equal to A, then X = A.
   and multiplying by X, then
   $X^2 = XA$ and subtracting A^2 gives
   $X^2 - A^2 = XA - A^2$ and factoring results in
   $(X - A)(X+A) = A(X - A)$ and dividing by (X-A) leaves
   X + A = A but by step 1 X = A, therefore
   A + A = A or dividing by A gives 2 = 1 but we know this can't be!!!

6. An 18th century rhyme from the Mother Goose Rhyme:

"As I was going to St. Ives
I met a man with seven wives.
Every wife had seven kids,
Every kid had seven cats,
Every cat had seven kits,
Kids, cats, kits, and wives,

How many persons were going to St. Ives?"

THE NATURE OF MATHEMATICS

Karl J. Smith.

Answers to Activity 2.1.2

1. $8 < x < 32$
2. a: 450 yds b 5/3
3. ABC-->BDC-->ADB The Order of the letters is the key to mapping.
4. 13.
5. EF is 12.5.
6. Can't divide by zero, see step 4.
7. Only one (read the first line). The assumption is the others were going in the opposite direction.

Write your summary

See some pictures in Appendix 5 for triangular geometric figures in buildings.

PLANE AND SOLID GEOMETRY ESSENTIALS

Chapter 2
Session 2
Types of Triangles

Mathematics is not a spectator sport!

Anonymous

Teacher: Ask what the quote means. Also, there are quite a number of theorems in this chapter. It is suggested that instead of showing the students how to prove, it is better to provide leading questions and guide the students through the proofs. It was stated by Frank Allen, a great teacher, that Geometry without proof is not Geometry, but Geometry with too much rigor only produces rigor mortise.

As you would expect some triangles, due to their use, have special names. This method of naming triangles is for the ease of communication. The common names for special triangles are **Isosceles, Equilateral, Scalene, Equiangular, Right, Acute,** and **Obtuse.** You see many cases of triangles in architecture if you look for them. How many triangles are in the following picture? What kind of triangles are they from the above list?

See pages 47 and 48.

Record in your notes the definitions for the seven special types of triangles for future reference. Use a dictionary if necessary! The lits includes, equilateral, isosceles, scalene, right, obtuse, similar, and congruen.

Comment: Most students have a fairly good idea what these definitions of the basic triangles are from their previous math. classes, but review these definitions and correct them if necessary.

Class Discovery Activity 2.2.1

Draw a larger figure (use a ruler) for each of the of the seven types of triangles and measure their angles with a protractor and write a definition for each. Three of the 7 types are below.

Examples:

The above could each also be classified as Isosceles!

What is your conclusion with regard to the angles for of the above triangles?

Do you understand the meanings for the names Right, Isosceles and Obtuse? (It is assumed you had these types of triangles in prior classes.) Follow the argument below to justify the angles opposite the equal sides are equal.

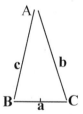

Label the angle vertices A. B. C. Label the opposite sides a, b, and c.

Given: **Isosceles Triangle ABC with AB = AC or c = b and D is the midpoint of side a.**

Can you justify that ∠B is congruent to ∠C and hence m∠B = m∠C?

(Hint: Draw segment AD as the bisector of BC and what do you then know about the two triangles?)

Teacher: Let the students explain their reasoning.

Label this conclusion as Theorem 6: (Write the theorem in If (the given) Then (the conclusion) form. (Also call it the Isosceles Triangle Theorem.)

Question: Do you think the angles in an equilateral triangle are all equal? Can you define what an equilateral triangle is? (Draw one.) Can you explain why the measure of all the angles in an equilateral triangle is the same? Can you justify that they are equal? (Justify your conclusion using the method from the above Isosceles case.) Class discussion and reasoning?

Label the conclusion as Equilateral Triangle Theorem 7.

The following question is easy to answer but the justification is harder. Do you think there is a relationship between the measure of the angles in a triangle and the opposite sides? What is your thought?

You may need to draw several larger triangles and measure the angles and sides. What do you think the relationship is? (see the figure below)

From the figures you drew, your guess is probably that the length of the sides corresponds to the size of the opposite angles, but can you justify the guess? Follow the Proof below!

Teacher: This proof may be a good class exercise so all will understand the conclusion.

In the above triangle:

Given: m∠B > m∠C > m∠A and you think side AC > side AB > side CB
(Recall that the symbol > from Algebra is read as "is greater than.)

Justification: (Follow this argument.)

We know m∠B > m∠C from the given, so at point B copy angle C so that angle CBP is formed and BC =PC.

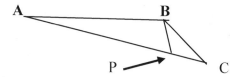

Triangle PCB is isosceles (why is BP=PC?).

AC + CB > AB. but PB = PC so AP + PC > AB. (Two sides of a triangle are greater than the third side, or the shortest distance between two points is a straight line segment.)

Since AP + PC = AC, therefore AC > AB.

Hence: The largest side is opposite the largest angle.

Comments: Students' feel this theorem should be an assumption or a definition.

Theorem 8: **In a scalene triangle, there is a correspondence between the length of the sides and the angles opposite the sides. The larger the angle, the larger the opposite side.**

Class Challenge: In a triangle, if the 3 angles are equal, then do you think the triangle is equilateral? Prove your answer.

Hint: Insert an altitude or an angle bisector.

Home Activity 2.2.2

The vertex angle in an isosceles triangle is the name given to the unequal angle.

Vertex angle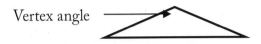

If the vertex angle is 50 degrees, then what is the measure of each of the other two angles? (Called the base angles.) Answer: 65 degrees

1. What are the measures of each side in an equilateral triangle, if the perimeter is 54 feet? Answer: 18 ft.

2. If the sides of a triangle are 23, 15, and x, then x is between what

    two values? Answer: 8 < x < 38

    Teacher: Listen to their explanations

3. Take and cut out five pieces of stiff cardboard 2 at 5 inches by1 inch, and 3 at 8 inches by 1 inch, put pins in the corners so 4 pieces and form a 5 by 8 polygon. This is a good home project or class project!

pins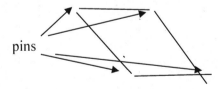

Is your figure flexible?

Teacher: This could also be done as a demonstration.
Now take the other 1 by 8 "stick" and flex the figure until the stick can form a diagonal. (The segment in figure on the next page forms 2 triangles)
Attach it with the pins. Is the figure still flexible?

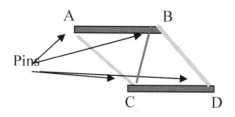

This is why the triangle is called a basic figure in construction. A triangle is a rigid figure!

4. Plot the following points on graph paper, connect the points, label the points and name the type of triangle that is formed.

    a. A(0,0) B(0,5) C(5,0) Name ?
    b. D(-5,0) B(0,7) C(5,0) Name ?
    c. D(-5,0) B(0,5) E(0,3) Name ?

Activity 2.2.3
A few Key Theorems

Definition 12: A parallelogram is a four-sided polygon with the opposite sides parallel.

Draw, using your ruler, three different shaped parallelograms and their diagonals.

2. Measure the opposite sides and also the two segments of the diagonals in each and write your conclusions. Do you think this is valid for all parallelograms?
You should have two conclusions by **inductive reasoning**. Try to prove each.

Given: Parallelogram ABCE with diagonals AC and BD
Prove: The point O, where the diagonals intersect, also bisects the diagonals. First draw a lager parallelogram and measure the segments AO and OC, DO and OB. What do you observe with respect to the AO and CO, DO and OB?

PLANE AND SOLID GEOMETRY ESSENTIALS

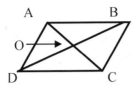

If you can justify that DO equals OB, then O bisects segment BD. Using the same method, prove that AO = OC. Is triangle ABC congruent to triangle ACD? Hint: Definitions are valid when reverse! See the definition of Parallelogram. Definition 3. Label this as:

Theorem 9a: **If the figure is a parallelogram, then the diagonals bisect each other.**

The following theorem should be proved in class.

Theorem 9b: **If the polygon is a parallelogram, then the opposite sides are equal and the opposite angles are equal.**

Definition 13: A rectangle is a parallelogram with right angles.

2 a. Draw a rectangle using your ruler and measure the diagonals. What is your conclusion about the diagonals? Prove your conclusion and label it **Theorem** 9c.

   Good class activity for better understanding.

   b. If in a parallelogram the diagonals are perpendicular, then it is a Square. Prove your answer and label as **Theorem** 9d.

   c. If the figure is a quadrilateral (4 sided) and the diagonals bisect each other, then the figure is parallelogram. Prove your answer and label it as **Theorem** 9e.

Thinking Question:

3. The manager for the school's baseball team each day counted the baseballs. He observed an odd result. If he counted them two at a time, he had one left over. If he counted them four at a time, he had one left over, and if he counted them 5 at a time, he had none left over. What is the smallest number of baseballs he counted. What is the next possible number of baseballs? (Hint: There are more than 10 balls and if divided by 2 then one was left, if divided by 4 then one was left, and if divide by 5 then none were left.) Tell the problem to your dad and both of you can try to solve it!

2. The converse of Theorem 9a: **If the diagonals of a 4 sided polygon bisect each other, then the polygon is a parallelogram.**
Label this Theorem 9f.

Write your Summary

Record the theorems to date and any additional ones your teacher may have suggested.

Conclusion: You also discovered that a triangle is a rigid figure and is called the **basic construction** figure.

Bridges need to be constructed with sturdy rigid figures which means lots of triangles. (Picture: Mount Hope Bridge in New England, about 1936.) See the next page.

Challenge

Draw a quadrilateral with the diagonals AD and CB, like the one below with AD and BC bisecting each other. Do you think the triangles are congruent? Do you think the figure is a parallelogram? Draw a few figures that fit the conditions above and then answer the question and justify your conclusion.

Teacher: Explain what the converse of a statement is.

This is the converse of Theorem 9a, label it **Theorem** 9f.

Question: How would you make the figure a rectangle? (Two possible answers.)

Extra Credit Activity

Students (extra credit) may volunteer to prepare a picture show of buildings in the community. illustrating the applications of geometric figures. Their phones make this an easy and interesting project. A few examples are:

LOOK THE TRIANGLES

From Elander's File
Triangles are rigid figures!

House illustrating applications of similar geometric figures.

See more pictures from the Elander File in Appendix 7.

Opportunity: Student photographers to create a set of pictures of geometric figures in buildings in their community or where they have visited.

PLANE AND SOLID GEOMETRY

Chapter 2
Session 3
What is Area?

Understanding evolves from work, appreciation from applications.

Unknown

A triangle divides a geometric plane into five sets of points or regions. In the figure below the sets of points are: the interior of the triangle (dark, set 1), the 3 line segments of the triangle (set 2), the exterior (set 3), the black rectangle border (set 4) and the white exterior of the page (set 5).

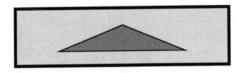

This session will determine ways to calculate the interior area of some geometric figures. These basic geometric figures are called polygons. Determining the measurement of these interior regions is used in everyday applications, such as landscaping, decorating, and building construction.

Definition 14: The AREA of a geometric figure is the number of square units contained in the interior of the figure.

In the figure, the term area refers to the red region or the interior of the square. Naturally, in the real world these problems involve figures of all shapes and sizes, polygons as well as curved figures like circles. The problem is to determine a method for calculating the number of square units the figure contains.

The selection of the proper square unit is important.

The following figure 1 is a square inch and figure 2 is a square centimeter.

Figure 1

Figure 2

The proper unit of area must be selected in order for the answer to have a meaningful interpretation. Consider a room that contains 100 square feet, but to state the 100 square feet as 14,400 square inches or as 92,903 square centimeters would provide no basic understanding. The selected unit of measure is very important for understanding.

Class Discovery
Activity 2.3.1

1. What is the area of each of the following figures?

Hint: Count the squares.

Figure 1:   Figure 2:

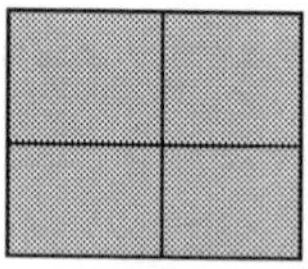

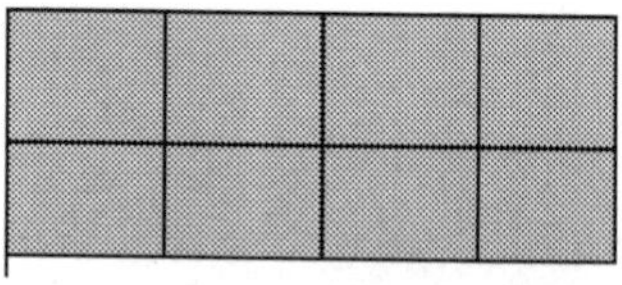

Figure 3:

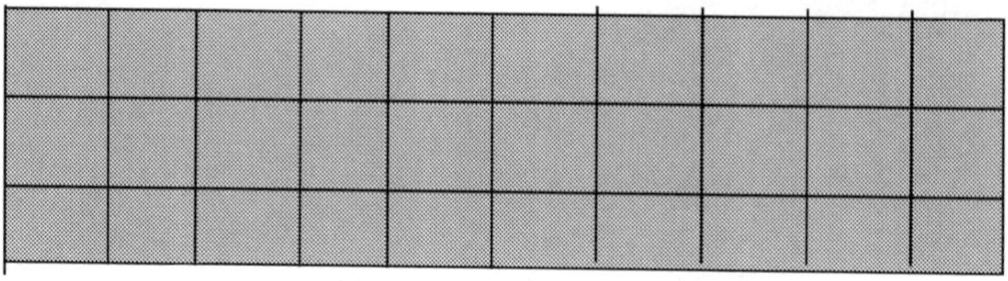

2. What are the values of the length and the width in figure 1?

Figure 2? Figure 3?

Complete the following table.

Figure	Length	Width	Area (counted)
1.	————	————	————
2.	————	————	————
3	————	————	————

3. What do you observe as to an easy way to calculate the area of these figures?

Since your observation, no doubt, is that the area of a rectangle is equal to the length times the width or base times height. Since this is not a proof, it will be called a postulate or an assumption to cover the general case. (Do you recall the definition of a postulate?)

Postulate   12: The area of a rectangle is equal to the length times the width and the answer is in square units. (The units of measure for base and height must be the same.)

The key word in this postulate is rectangle and from the definition you know that a rectangle is a parallelogram with right angles.

Where AB is parallel to DC, AD is parallel to BC, and all the angles are right angles (each measure 90 degrees) this parallelogram is called a rectangle. The figure above will be used to prove the area of a right triangle. Follow these steps carefully and ask your questions if needed for understanding.

Class Discovery
Activity 2.3.2

1. The area of a rectangle is the measure of the base times the measure of the height or DC times AD in the figure. (Postulate 12)

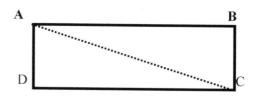

Draw the above figure and the diagonal AC. Why can this be

a. Why can this be done?   Two points determine a line segment.

c. The triangles ADC and ABC are similar. Why?   AA

d. The triangles ADC and ABC are congruent. Why? Ratio of sides is 1.

e. This also proves AD = BC and AB = DC. Why?

f. Therefore, the area of triangle ADC is ½ the area of the rectangle. Why?

g. Therefore, the area of a right triangle ADC is (½)bh sq. units, where b is the base (DC) and h is the height (AD). Q.E.D. What does Q.E.D. mean? (See a dictionary)

Theorem 10: **The area of a right triangle is ½ the base times the height. (The base and height must be in the same units of measure.)**

2. You are probably thinking, what if the triangle is not a right triangle? Example: Area of triangle ABC using the formula is A = (½)bh.

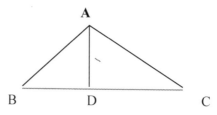

Method: The procedure is to convert the problem back to a former one. (Learn the new in the light of the old.) In the above case draw a perpendicular line segment from A to BC as shown above. Assume DB = 5ft., DC =12 ft., and AD = 6 ft., then what is the area of triangle ABC? (Ans. 51 sq. ft.)

Theorem 10: Stated that the area of triangle is (1/2)bh square units. It is a natural question to ask. What is the area of a parallelogram?

On your paper draw the following parallelogram with line segments AK and DE perpendicular to BC extended. Why can this be done?

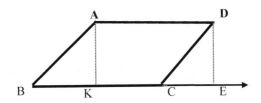

a. Segment AK is perpendicular to BC. See postulate 5:
b. Draw segments DE perpendicular to BC extended.
c. AK equals DE. Why?
d. Area of AKED equals KE units times AK units. Why?
e. Triangles ABK and CDE are congruent. Why? Triangles are Similar and ratio of sides is 1.
f. BK equals CE. Why? Corresponding Segments-congruent triangles
g. BC equals KE. Why? Algebra
h. Parallelogram ABCD equals rectangle AKED in area or BC times AK. Why? Algebra
i. Triangle ABC is congruent to triangle ACD. Why?

Theorem 11: The area of a parallelogram is equal to base times height or base times altitude. Formula: $A_{\square}$= bh sq. units

Comment: One method for determining the area of figures or polygons that don't fit the case of a triangle or parallelogram is to divide the polygon into triangles.

The following will illustrate how the above theorems are used.

Discovery Activity 2.3.2
Trapezoid

Draw on your paper a trapezoid.

Definition 15: A trapezoid is a four-sided polygon with only two sides parallel. See the figure below.

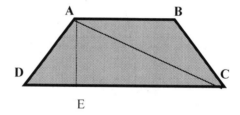

2. The trapezoid is divided into two triangles by segment AC, therefore the area is the sum of the areas of the two triangles. AE is the altitude for both triangles. Area of Triangle ADC + Area of Triangle ABC = ½ (DC)(AE) + ½(AB)(AE) or by using algebra the Area of trapezoid equals ½(AE) (DC + AB) square units.

Theorem 12: The area of a trapezoid is the sum of the two bases times ½ the measure of the altitude (h).

Formula: $A \text{ (trap)} = \frac{1}{2}(h)(b_1 + b_2)$ sq. units

Chapter 2: Applications
Activity 2.3.3

(Some answers are given or are in the summary so students can check their answers.)

1. A high school student has a garden in the shape of a trapezoid with these measurements: bases of 20 feet and 14 feet with the to other sides each 5 feet. The height of the trapezoid is 4 feet.

   a. Draw the figure and label it with the dimensions.
   b. What is the area of the total garden? 68 sq. ft.
   c. What is the area of the rectangle? 56 sq. ft
   d. What is the perimeter of the garden? 44 ft.
   e. Are the triangles similar? If so, why? SSS or SAS
   f. Are the triangles congruent? If so, why? Ratio is 1.

2. A rectangular park measures 75 yards by 40 yards with a two-yard wide walk around the perimeter. (Draw a figure to represent the grass area and the walk.)

   a. What is the area allotted for grass? 3000 Sq. Yds.
   b. What is the area of the walk? 476 sq yds
   c. What is the length of the total perimeter? (Grass and walk.)
                                                         246
   d. If the walk is used as an exercise walk and you walk around the perimeter 8 times and using the length from part c for the perimeter, then would you have walked more or less than a mile? Justify your answer. (How many feet in a mile? See appendix 4) **(Park perimeter is 5903 feet which is more than a mile.)**

3. If a map has a scale of 1 inch equals 2 miles, then a 3 by 5 inch rectangle on the map will represent how many square miles?

(60 sq mi.)

Class Discovery
Activity 2.3.4

1. Draw three different shaped triangles and using your ruler, draw the perpendicular bisectors of each side. What do you observe? Answer: The perpendicular bisectors of the sides intersect in a common point or are concurrent. Can you prove it for the general case? The following may help?

Method: Step 1. Can you prove that any point (P) on the perpendicular bisector of line segment AB is equidistant from the end points of the segment?

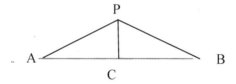

Step 2. Draw segments AP and BP.
Step 3. AC = BC, angles ACP and BCP are equal, and PC=PC, hence the triangles are congruent and AP equals BP.

Theorem 13. **Any point on the perpendicular bisector of a line segment is equal distant from the endpoints of the segment. Draw the figure.**

Challenge: Prove that if point P is equal distant from the end points A and B of a line segment, then P is on the perpendicular bisector. (Teacher: This is the converse of Theorem 13 and will be labeled Theorem 14.) Define converse, if needed again by some students.

Draw the figure and label it.

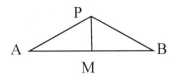

Given: AP = BP, and M is midpoint of AB
Prove: PM is the perpendicular bisector of AB

Related problem: Given angle ABC with its angle bisector labeled BD. Prove that if a point is on the angle bisector, then the point is equal distance from the sides of the angle.
(Hint: ASA). (Let the students show their proof) Label it **Theorem** 15.

Now prove or justify the converse of **Theorem** 15 which is: If a point is equal distance from the sides of an angle then the point is on the angle bisector. Label it Theorem 16 .(Hint: SAS) Write the theorems 15 and 16 in if-then form! They are needed in the next session!

CHALLENGE
(Students may need hints or additional help!)
Week end problem

Thinking Question: The number of possible scalene triangles having a perimeter of **less than 13** is? The sides are whole numbers! Hint: A+B > C
(One case for the sides is 3,2,4.)
Which of these is the answer?
a. 6   b. 3   c. 5   d. 4   e. more?

Summary

Record your definitions for: Triangle, polygon, parallelogram, rectangle and trapezoid, area, altitude, perimeter, and square units were introduced.

Record the Theorems in your notes: Write out the polygon theorems in If-Then form.

Theorem 10: The area of a triangle is ½ the base times the height or ½ base times the altitude. Formula: $A_\triangle = (½)b)h$ sq. units.
Formula: $A_\triangle = (½)bh$ sq. units

Theorem 11: The area of a Parallelogram is equal to the length times the width and the answer is in square units. (The basic unit of measure for base and height must be the same.)
Formula: $A_\square = bh$ sq. units

Record Theorems 12, 13, 14, 15, and 16.

Research and report:

Do a computer search or use your library to investigate:

a. History of the English System of measurement
b. History of the Metric or SI system of measurement. (A student or a committee may wish to give a report on the history of SI. (This will make future homework more meaningful.)

Interesting challenge Problem and conclusion
(Your parents may enjoy this?)

Conclusions are made after collecting information from as many creditable sources one can find. In today's world these sources are TV shows, magazines, books, lectures, radio reports and experts (many shows consist of pseudo experts) plus friendly discussions.

Number Theory, it has been said, not only turns many students on to math, but also increases their thinking ability.

The following types of problems require data organization, observation, plus discovering a possible conclusion and then testing their discovery. One of these motivating types is the problem of determining the sum of the first odd 40 counting numbers? (It is said that Gauss, one of the greatest Mathematicians when in grade school answered the question for the sum of the first 100 counting in a few minutes.)

The method is to record a few simple cases as follows.

Cases	Numbers	Sum
1	1	1
2	1+3 =	4
3	1+3+5 =	?
4	1+3+5+7 =	?
N	1+3+5+7...40...N =	?

Do you see any pattern? The first two add to 4. The first 3 odd numbers add to 9. What do you predict the first 5 add to? Do you see the pattern?

Continue the table until you see an easy way to arrive at method for the sum of the first N odd counting numbers. Remember you are looking for an easy method to give the sum of the first odd 40 counting numbers or integers. The answer is: 1600

Suggestion: The first one in the class to see the general method becomes the class Gauss for the day. He or she will never forget it. What is the sum of the first N odd counting numbers? N^2

PLANE AND SOLID GEOMETRY ESSENTIALS

Chapter 2
Session 4
Pythagorean Theorem

Neglect of mathematics works injury to all knowledge.

Roger Bacon

This theorem is said to be the most important theorem in Geometry. In other words, this is the one to fix in your memory. It is on every test related to Geometry. That alone tells you it will be on your college or tech school entrance exams. An example of the theorem is below and a proof will be provided later. The 3-4-5 case was known over 4000 years ago.

The theorem pertains to a right triangle and the relationship of the sides. Draw a right triangle and label it ABC with C the vertex of the right angle.

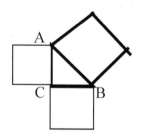

Label the sides of the triangle, a, b, and c with "a" opposite vertex A, "b" opposite vertex B, and "c" opposite vertex C.

Construct your drawing so that side "a" is 4 cm, "b" is 3 cm, and side "c" is 5 cm. Euclid pictured it this way. Notice: 3 squared (9) plus 4 squared (16) equals 25, which is equal to the area of the large square (25).

Which means, in this case, the square on the longest side of a right, **hypotenuse,** triangle is equal to the sum of the squares on the two other sides, or the sum of the areas of the squares on the sides equals the area of the square on the side opposite the right angle. Question: Do you think it is true for all right triangles?

Definition 16: In a right triangle the side opposite the right angle is named the hypotenuse.

Discovery Activity 2.4.1
Complete the values for the ? marks.

Case	a	b	c	$a^2 + b^2$	$= c^2$
1	3	4	5	9 + 16	= 25
2	5	12	?	25 + 144	= ?
3	7	24	?	49 + ?	= ?
4	9	?	?		
5	11	?	?		

Can you (the student) complete the missing values? Use your calculator.

Discovery a few more cases? Use your calculator. Look for a pattern. If you do find others cases, then report the cases to your teacher, fellow students and your parents.

This illustrates the Pythagorean Theorem. Can you state the theorem in if-then form?

There have been over 300 hundred proofs of this theorem, one by a former President of the United States, and one by a high school student in the late 1930s. Special cases were known over 4000 years ago. This topic, Pythagorean, provides an opportunity for an interesting report and presentation.

Notice: We have not yet proved the theorem!

Research Source: THE PYTHAGOREAN PROPOSITION by E. Loomis or any book on the History of Mathematics.

One of the over 300 proofs is explained below. Provide a reason for each step. The teacher may help or give hints to the students.

Given a square with side S.
What is the area of the square?
Answer: $A = s^2$ sq. units or $A = (a + b)^2$ sq. units.

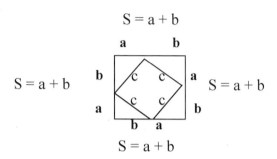

$$S = a + b$$

Is the following true?
Area of the large Square, which is S^2, is also equal to the sum of the area of 4 triangles plus the inner square. therefore the following is valid.

$(a+b)^2 = (4)(1/2)ab + c^2$ (c is the side of the inner square side length)

Simplifying:

$$a^2 + 2ab + b^2 = 4(1/2)ab + c^2 \text{ by algebra}$$

Hence: $a^2 + b^2 = c^2$ by algebra

Theorem. 17: **If given a right triangle with sides a, b, and c the hypotenuse, then $a^2 + b^2 = c^2$.**

Comment: As stated, there are over 300 proofs for this theorem. Two other proofs will be shown later on.

Class Activity 2.4.2
(Some answers are at end of the Activity.)

1. Which of the following are right triangles? Explain your answers. (The sides listed in a, b, c order. Draw and label a triangle for each problem. Use your calculator when needed.)

   All are right triangles.

   a. 6, 8, 10   b. 1.5, 2, 2.5   c. 5, 12, 13   d. 2.5, 6, 6.5
   e. 9,40, 41   f. n, (n-1),   Ans. $\sqrt{2n^2 - 2n + 1}$

2. Given: Equilateral triangle, ABC, with side AC equal to s and segment CD is perpendicular to side AB.

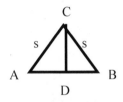

   a. What is the length of AD (in terms of s, the side)?
      Answer: s(√3)/2 How do you know? (Hint: If AC=s and AD = s times 1/2)
      Students will probably need some help with this class activity.

   b. What is the measure of angles ACB and ACD? Why?
      Answer: 60 and 30 degrees (Be able to justify your answers.)

   c. If each side is 6 units, then what is the measure of AD? Why?
   d. If each side is 6 units, then what is the measure of CD? Why?
   e. Could triangle ACD be called a 30-60 right triangle? Why?
   f. Write the results from a and b as **Theorems 18 and 19.**

Answers to Activity 2.4.2

2. a. 3√2 units b. 60°, 30° c. 3 units d. 4.24 units e. yes f. see s below.

Theorem 18: **In a 30-60 degree right triangle the side opposite the 30-degree angle is ½ the hypotenuse. S/2**

Theorem 19: **In 30-60 degree right triangle the side opposite the 60-degree angle is half the hypotenuse times the √3. s(√3)/2**

Definition 17: In a triangle, an altitude is defined as the line segment from an angle vertex perpendicular to the opposite side. (The side may have to be extended.) Draw the altitude from A.

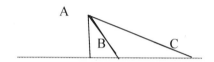

Class Activity 2.4.3
(Teacher may have to guide the students on these Theorems Try to anticipate their questions!)

1. Draw three triangles (acute, obtuse, right) and draw the three altitudes in each case.

   a. What do you observe? (Use your ruler.)
   b. Write a conclusion. **Altitudes seem to intersect at one point.**
   c. **A proof that the 3 altitudes of a triangle are concurrent follows.**

   Hints for the proof.
   Draw a triangle ABC with three altitudes and P is the point where they intersect.
   See Theorems 13 to 16.
   AD is altitude to BC. (p is on AB and p is on CD and p is on BE, you think!)

3. What does problem 2 mean as to the point where the altitudes intersect? P(the point of intersection)

   Explain that this is a point of concurrency. Call this **Theorem 20.**

4. Prove: If a point is on the angle bisector, then it is equidistant from the sides of the angle. See the angle bisector theorem.

5. Draw the figure like the one below with the three angle bisectors.

   Label the point where the 3 bisectors intersect as O. (See Theorems 15 and 16.)
   From Theorems 15 and 16, what do you know that AO = BO and also CO?

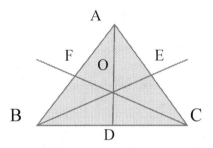

This also says BO = CO, therefore O is the point of concurrency for these two segments! This will be **Theorem 20.** State this in if- then from. The following may help. Draw a triangle and put in the three altitudes and notice the altitudes appear to intersect at 0. Question: Wonder if it can be justified? The ancient mathematicians wondered also.

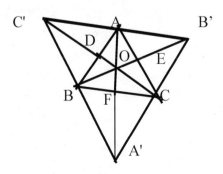

On your diagram are the following.

Through each vertex (ABC) draw a line parallel to the opposite segment so that A'C' is parallel to AC, A'B' is parallel to AB, and B'C' is parallel to BC. (see above figure) Therefore, the diagonals of the parallelograms, ABCB', AB'CB, ACBC' intersect at point O. (See Theorem 16) Also the point 0 is equidistant from B' and C' and also 0 is equidistant from points C' and A', also O is equidistant from points A' and B'.

a. Hence 0 is the point of concurrency.
   (Try the case where the triangle is obtuse. Extra credit problem for a student or a group of students.)

Another way stating this: If two altitudes intersect at P so that AP = BP and a third altitude intersects such that BP = CP, then P is a common point, or the lines are concurrent.

Theorem 20: The three altitudes of a triangle are concent.

Definition 18: The median of a triangle is defined as a line segment from the vertex to the midpoint of the opposite side.

Class Discovery Activity 2.4.4

Draw two sets each of different triangles (acute, obtuse. equilateral) and then draw in each set the 3 angle bisectors. What is your conclusion from each set? This proof for the concurrency of the angle bisectors is below, but the proof relating the medians will be justified in another session.
Hints as to the proof that the angle bisectors are concurrent.

1. The angle bisectors of a triangle, we assume intersect at a point (concurrent) from the following:

   a. Draw the following triangle ABC as below.
   b. Color the angle bisectors red and label them AE, BD, CF.

c. Label you think is the point of concurrency, O.
d. Does the point O appear to be the same distance from the 3 sides of the triangle?

Angle ABC with the angle bisectors, BD, AE and CF and any point on the angle bisector is equal distant from the rays of the angle. (Th.13,14,15)

AE, CF and BD
are angle bisectors

(See the **Theorems** 13,14 and 15 which are used in the proof also.)

Does it appear that any point on the angle bisector is equal distance from the two rays, in other words the rays BD and AE intersect at O and the rays CF and BD intersect at O since O is equal distance from AB, AC, and BC. Can you prove that if a point is equal distance from the sides of a triangle, then this concurrency point is on the angle bisector. Do you think that if a point is on the angle bisector, then the point is equidistant from the rays of the angle? If your answer is yes, then write the **theorem** and label it as **Theorem** 21. Hence, F0 equals E0 and D0 and the bisectors are concurrent.

There are three medians in a triangle also, so draw a triangle with the three medians and what do you observe? Do you think they are concurrent? **This will be justified in another section.**

GEOMETRY ESSENTIALS
FOR
P LANE AND SOLID
Chapter 2
Session 5
Review

Review: Short answer questions-use ruler for drawings.
Suggestion: Do not check your answers until you complete the review.

1. Are the following valid definitions? Be able to defend your answer.

    (Remember: A definition is valid, if the definition is true when reversed.)

    a. A restaurant is a place that serves food.
    b. Mathematics is a useful course.
    c. A postulate is an assumption.
    d. A triangle is a set of three non-collinear points and the line segments determined by the three points.

2. What are the three geometric terms that are classified as undefined?

3. How many points are needed to determine a geometric line?

4. The points on a line correspond to the numbers on the _____ _____ line.

5. Draw a ray and label it AB.

6. Draw a line segment and label it AC.

7. Draw a triangle and label it CDE.

8. A geometric plane is determined by ___ ___ ____ points.

9. What is the sum of the angles in a plane triangle?

10. How is the distance between points A and B determined?

11. What is a theorem?

12. When are triangles similar?

   a.   b.   c.

13. When are triangles congruent?

14. a.   What are parallel lines?
    b.   What are skew lines? (Need a dictionary?)

15. Draw three sets of 3 triangles (acute, obtuse, scalene triangles) and label each ABC. Use your ruler and protractor.

   a. In one draw the medians.
   b. In another set draw the three altitudes.
   c. In the third set draw the three angle bisectors.
   d. Write three observations.

16. The shortest distance from a point to a line is the _____ distance.

17. If A implies B and A is given, then _____.

18. Draw a rhombus.

Answer for Activity 2.5.1

1. C and D are definitions.
2. Point, line and plane
3. 2
4. Real number line
5. A ⟶ B

6. A ___ B

7. B ◁▱▷ C (with A above)

8. 3 non-collinear points
9. 180 degrees
10. Use a ruler
11. A math statement that has been proved.
12. Angles that are equal and the correspondence of the sides in same ratio.
13. Similar figures and the ratio of the corresponding is 1,
14. a. Lines on a plane that do not intersect.
    b. Line not on the same plane and do not intersect. Conclusion: The altitudes intersect at a point, the angle bisectors intersect at a point, and the medians intersect at a point, but the point where the medians interest needs to be justified.

15. Perpendicular Straight line
17. B
18.

Record your errors for more review!

PLANE AND SOLID GEOMETRY ESSENTIALS

Chapter 2
Session 6
Applications

You cannot fake in mathematics, no one can be fooled.
You can either prove (solve)... or you cannot.

Jerry P. King
THE ART OF MATHEMATICS

Suggestion: Use your notes and other tools, organize your work and be able to justify your answers. Some answers are at the end of the activity. Units are not assigned to the answers and student will have to add them.

Activity 2.6.1
Applications

1. Equilateral triangle related questions:

   a. Draw an equilateral triangle with sides measuring 2 inches. Use your ruler and protractor. Label the vertices's A, B, C.
   b. Calculate the perimeter.
   c. Calculate the area. Hint: $2\sqrt{3}$ or 3.46
   d. Add the segments for the 3 medians to your drawing in 1a. Label them AE, BD, and CF.

2. In the figure for #1, label the point of intersection in "1d" as G. What is the number of degrees in angle AGB?   120°

   a. What is the area of triangle AGC? $\sqrt{3}/3$ q inches
   b. What is the length of segment ED?   1 inch

c. Calculate the length of segment BE?   Ans: $\sqrt{3}$
d. What is the physical property of point G? Hint: Center of_____.
e. What is the measurement of angle ADB? 90°

3. A city park is triangular in shape with sides measuring 900, 800, and 700, all in feet, but for easier computations just use 9, 8 and 7 and multiply the answer by 100. The mayor asked the teacher if the class could to calculate the area? The class's response was, "yes sir". The mayor requested the answer by the next day, with a drawing, and the explanation as the solution.

This is a good class exercise with the teacher working the problem on the board and the class asking questions.

Hint: Draw Triangle ABC label the sides a, b, and c and solve for h. Using the given information with the sides equal to 900, 800 and 700. Draw an altitude from one of the vertices to the base and label one segment of the base as x, then write two equations involving x and h. Solve for h and then solve for the area using the triangle formula.

Hint: $7^2 = h^2 + x^2$ and $8^2 = h^2 + (9 - x)^2$

Now solve for h!
Answer will vary depending on rounding off values.

4. A park is in the shape of an isosceles triangle with the equal sides each measuring 150 feet and the equal angles are each 30 degrees. What is the length of the base to nearest foot? Hint: Draw the figure and insert the altitude. Base is 260 ft.

5. If a square has sides measuring 10 meters, then what is the length of each diagonal? Diagonal = 14 meters

6. The school principal needs to know the height of the school's flagpole.

   You are given the following facts: From a point 75 feet from the base of the pole two students observe the following.

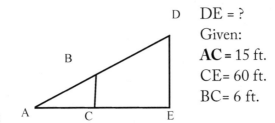

   Ground level:

   DE = ?
   Given:
   AC = 15 ft.
   CE = 60 ft.
   BC = 6 ft.

   Hint: Use similar triangles.
   What is the height of the pole to the nearest foot? H = 36 ft.

7. Given: If you are an insurance agent, then you should have studied Mathematics. (This is a true statement and valid. A → M)

   a. Write the converse of the given and indicate if it is true and valid.
   b. Write the inverse of the given and indicate if it is true and valid.
   c. Write the contrapositive of the given and indicate if it is true and valid.

8. Induction: A person plans a garden and has 120 feet of fencing. What is the length and width of the largest rectangle (area) that can be fenced? In other words, the length times the width is to be a maximum.

   Hint: 2L + 2W equals 120 and LW = a Maximum?

	Length	Width	Area
Try	5	55	275
Try	10	50	?
Try	15	?	?

   Hint: Continue changing lengths until you arrive at a conclusion for the maximum area. Write your conclusion.

9. From your solution to #8, what would you predict the maximum area, if the amount of fencing is 360 feet. Hint: 2l + 2w = 360 now use the method in #8. (8100 sq feet.)

10. Draw a 30 – 60 degree right triangle. Label the vertices A, B, and C, where C is the vertex of the right angle and angle A is 60 degrees. The hypotenuse is 10 units. Hint: Theorems 15 and 16.

    a. Draw the altitude from C and label it CD.
    b. How many triangles are formed?
    c. If similar, map the triangles and indicate the size of each angle.
    d. If the measure of the hypotenuse is 10, then what is the measure of the sides in the three triangles?

11. Everyday thinking type problems

    Why may a four-legged table wobble?

    Using the forms of implications to arrive at conclusions.
    Remember: If A --> B Statement (Valid)
              If B--> A Converse (Not necessarily valid)
              If not A --> not B. (Inverse, not necessarily valid)
              If not B --> not A (Contrapositive, valid)

    Now apply the above to the following and indicate which you think are True or false and which are Valid or invalid?

12. Let A be Successful and B represent you are not. Write each below in A-->B form.

    a. If you are Successful, then you are not Lazy. Hint: A --> B
    b. If you are not successful, then you are lazy.
    c. If you are lazy, then you are not successful.

13. Let F be "fun to be with" and H be "happy friends"

If you are fun to be with, then you have happy friends.
(F --> H) Assume this is valid.

a. Translate: If H then F. Is it always valid? Why?
b. Translate: If not F, then not. Is it always valid? Why?
c. Translate: If not H, then not F. Is it always valid? Why?

14. Statement: If you are a cowboy, then you wear handmade boots.

   (C-->B)
Given: Tex is wearing handmade boots.
Which of the following is a valid conclusion and why?

a. If you wear handmade boots, then you are a cowboy.
b. If you are not cowboy, then you don't wear handmade boots.
c. Tex is not wearing handmade boots, then he is not a cowboy.

Hint: Write the converse, the inverse, and the contrapositive of statement.
      What can you say about Tex?

Teacher: Objective for exercises 11-13 is to point out the validity of
         the forms of an implication. These types are used in everyday
         thinking, mostly without the understanding of each. Recall:
         Statements are true or false and conclusions are valid or invalid.

Answers: Activity 2.6.1

1. a. Drawing

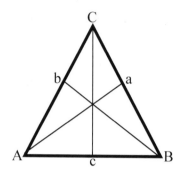

Hint: **It may help if you label side AC as b, side BC as a, and c is side AB**

b. P = 6 inches   c. 1.732 sq. in.   d. Drawing   e. Drawing
f. 120 degrees   g. 2√3/3 sq. in.   h. 1 in.   i. 3
j. Center of gravity   k. 90 degrees

2. See problem.
3. See problem.
4. 30 feet to the nearest foot.
5. 14 ft or (10√2)
7. Not always T and invalid   b. Not always T and Invalid   c. Valid and true
      M-->A   not A-->not M   not M-->not A
8. 30 ft by 30 ft Area = 900 sq. ft.   9. 90 by 90 ft. Area = 8100 sq. ft.
Draw the triangle and label.
Angle measurement:

A = 60° B = 90° C = 30° ABD = 30° CBD = 60° BDC = 90°
AB =? BC =? BD =? AD =? BC =? AC =?

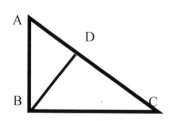

Measurements (Give answers to 2 decimal places)

AB = 10.00   AC = 5.00   BC = 8.66   AD = 2.50

CD = 4.33   BD = 7.50

10. Four points can determine four planes. ABC, ABD, BCD, ACD

11. a. not valid (Converse If B → A)

    b. not valid (Inverse- not A → then not B)

    c. Valid (Contrapositive- not B → not A)

    (Remember validity does not mean the statement is true or false.)

12, 13. a. not always (Converse)

    b. not always (Inverse)

    c. valid (Contrapositive)

14.   c. Tex may be a cowboy (converse may be valid.)

    Diagram also explains it, since other people can wear hand-made boots other than cowboys.

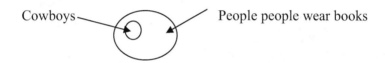

Cowboys   People people wear books

Assume Tex is a cowboy from his name and is wearing handmade boots.

Teaching Suggestion: Carefully explain the missed problems in class or let students explain their method. Understanding of the Decision-Making problems is very important.

Class Discovery Activity 2.6.2

Conclusions are made after collecting information from as many credible sources one can find. In today's world these sources are TV shows, magazines, books, lectures, radio reports and experts (many shows consist of pseudo experts) plus friendly discussions.

Number Theory, it has been said, not only turns on many students to math, but also increases their thinking ability. The following types of problems require data organization, observation, plus discovering a possible conclusion and then testing their discovery.

One of these motivating types is the problem of determining the sum of the first100 counting numbers. (As stated before, it is said that Gauss, one of the greatest Mathematicians, in grade school answered the question in a few minutes.)

The following (hint) is one approach to the solution and illustrates that organizing information can help to discover the solution method. Arrange your work in the following manner for an easier way to see the method. Do you see the relationship between the N-number in the cases and the number in the sum?

Cases (N)	Numbers	Sum
1	1	1
2	1+2 =	3
3	1+2+3 =	6
4	1+2+3+4 =	10

Do see an easy way to figure Sum given N?
Can you calculate the sum of the first 100 positive integers?
Answer: 5050
The answer for the sum of the first N integers is:
Sum of 1 to N = n(n+1)/2

Now a method to solve for the sum of the first N even numbers.
Question: What is the sum of the first 40 even numbers?
Make a table like the one above and look for the pattern and easy way to solve for the answer.

Cases	Numbers	Sum
1	2	2
2	2,4	6
3	2,4,6	12
4	2,4,6,8	?

Continue the cases until you see the relationship between the case number and the Sum number or the answer.

Remember you are looking for an easy method to give the sum of the first 40 even numbers.

Answer is 1640.

Teacher Suggestion: The first one in the class to see the method becomes the class Gauss. He or she will never forget it.

Your parents will be very proof of you when you are able to give the answers for these sums!

Another interesting Take home Family problem.
(Ask your parents the following question.)

How many times you can fold a 11 by 8 sheet or any size sheet of paper in half? Guess first and then actually count as you fold!

Example:

Folds   0   1   2   3

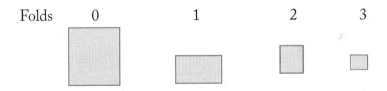

Try it with several types of paper! You may want to change your guess? Your parents will be surprised also!

Home and/or class Activity 2.6.3
An Interesting **Theorem**

The following activity is related to **Theorems** 11 and 12. The result is a very useful theorem, but not an easy theorem to prove.

Draw 3 triangles, one equilateral, one isosceles and one obtuse. The longest sides should be about 10 cm. In each triangle draw the three medians, use your ruler and label the point of intersection D. Estimate the length of each median and then measure the segment from each vertex to the point of intersection. the nearest tenth of a cm. and complete the following:

Triangle	Label	Lengths					
Equilateral	ABC	AB = ?	AC = ?	BC = ?	AD = ?	BD = ?	CD = ?
Isosceles	ABC	AB = ?	AC = ?	BC = ?	AD = ?	BD = ?	CD = ?
Obtuse	ABC	AB = ?	AC = ?	BC = ?	AD = ?	BD = ?	CD = ?

What does the distance to the point of concurrency from the vertex of each angle appear to be compared to the length of each median. Write your assumption.

Now back to the medians concurrency problem.

The three medians in the triangle below intersect at point O and the point is 2/3 the measurment of the median from each vertex of the angle. See drawing below. The original triangle is ABC with D, M, F the midpoints of the sides. The ojective is to **prove** the point O is the point where the medians intersect and the point O is 2/3 the distance of the median from from each vertex, CO is 2/3 of CF, and A0 is 2/3 AM, and BO is 2/3 of BD.

Teacher: This is suggested as a class activity with the teacher leading the discussion. It will review a number of theorems and there should be quite a few questions if not, the teacher should ask the questions. The figure below will be used to help prove the theorem.

Given: Triangle ABC with M, D and F midpoints of the sides and the Medians are AD, BM and CF are medians intersecting at 0

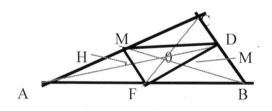

1. Draw line segment DM is it parallel to AB? Why?
2. Draw line segment MF is it parallel to CB?. Why?
3. Polygon MFBD is a parallelogram. Why?
4. Diagonal MB and DF bisect each other, hence GD =GF, BG= MG.
5. But MDF are the midpoints of the sides and HC, BC and AB.
6. Therefore, O divides BM is at the 2/3 point and O is at the 2/3 point for AD and also CF.

Theorem 22: The medians of a triangle intersect, or are concurrent, at a point that is 2/3 the distance of the median from each vertex. (Very useful theorem!)

Another proof for Theorem 12 (Pythagorean Theorem)

Recall it was stated that there are over 300 proofs of the Pythagoren theorem. Here is another one which you can prove, (Good extra credit problem with student's explanation)

Given: Right triangle ABC with altitude CD with sides of a, b and c.. **Map** the 3 triangles and write the ratios as indicated below.

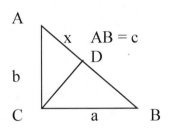

$AD = X$
$DB = y$ or $AB = x + y$
$AB = x + y$
$CB = a$
$AB = c$

The Mapped three triangles.

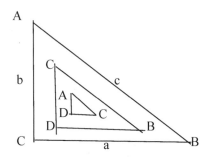

(You may want to make a larger drawing.)
The objective is to justify that $a^2 + b^2 = c^2$

From similar triangles you can write the following and using $AD = x$, $DC = y$, the following can be written

$b/x = c/b$ ---> $b^2 = cx$
$a/y = c/a$ ---> $a^2 = cy$
and adding results is $a^2 + b^2 = cx + cy$, but $cx + cy$ is equal to $c(x + y)$ which is c^2 and $a^2 + b^2 = c^2$

Therefore QED!

Teacher: The students will need help on the last few steps!

(The **hint** for this method was found by the author in an 1890s Geometry text!)

Write your Summary

In your note book copy the theorems in If-then forms with examples. **Try to find the center of gravity for the following objects.**

3. A baseball bat. Ask a coach about the use of the center of gravity.
4. A yard ruler.
5. A piece of cardboard.
6. A pencil, or a spoon.
7. A card board triangle. Locate the medians point of intersection.
8. A golf club, say the driver.
9. Discuss the above with your friends and also your parents.

Challenge Problem

Given: A trapezoid with the measurements AB = 15, DC = 9, and AB is parallel to DC with an altitude of 4. If AC and BD are is extended until they intersect at E, then what is the area of triangle CDE?

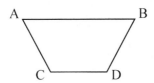

PLANE AND SOLID GEOMETRY ESSENTIALS

Chapter 3
Session 1
Circles

That they (all citizens) might excel in public discussions on philosophic or scientific questions, they must be educated which included rhetoric, philosophy, mathematics, and astronomy.

The Athenian Sophist School Curriculum
(480 B.C.E.)
F. Cajorie
History of Elementary Mathematics

You now have studied and reviewed the concept of area, but only for plane polygons which are figures consisting of line segments. The following definition of a polygon was stated in Chapter 1 Session 4.

Definition 9: A polygon is a plane closed figure of line segments determine by n points. Can you name the polygons?

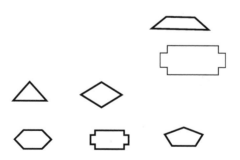

Not only do you need to be able to calculate the area of the above, but also you need to determine the sum of the interior angles, which is on many tests. Naturally, you would have to know the measurements to determine the area. To calculate the area you could divide the figure into triangles, which illustrates a useful procedure, time consuming! To **solve a new problem try to reduce the problem to an old problem.**

What if the plane figures do not consist of line segments or are a combination of line segments and curves, like the following?

The above figures could be, a circle, an ellipse, part of a circle, or like the sweep of a windshield wiper. In the real world the area of curved figures requires the use of mathematics for financial reasons as well as construction. These curves are related to a circle, so the circle is where we will start.

Definition 19: A circle is defined as a plane figure consisting of set of all points on the plane that are a given distance from a given point on the same plane. (The given point is labeled the center and the given distance is called the radius.)

Class Discovery Activity 3.1.1

The following activity will refresh your memory as to the terms related to a circle. Draw or trace a large circle on your paper, draw and identify (label) the following items in red on your circle.

Definition 20: Circle terminology. Insert in your circle the following items.
   · The given point in the above definition is called the center.
   a. Label it O.
   b. The line segment from the given point (center) to a point on the circle is called the radius. Label it OA.
   c. The distance around the circle is the circumference. Highlight it.

d. The line segment from one point on the circle to another point on the circle is called a chord. Label it BD.

e. If a chord contains the center point, it is called a diameter. Label it EF.

f. A line that intersects a circle in only one point is called a tangent.

Label it GT, where T is the point of tangency.

g. A secant is a line or a segment that intersects a circle in two points.

Label it MN where M and N are points on the circle.

h. A portion (segment) of a circle is called an arc. Label the endpoints of the arc RS and color or high light the arc.

Question: Do each of the above meet the condition of reversibility for a valid definition?

PLANE AND SOLID GEOMETRY ESSENTIALS

Chapter 3
Session 2
Circle Postulates

The ancient Babylonians (about 2000 BC) and probably others, such as the Chinese, considered the same problem related to the circle and solved the problem to meet their needs. The problem, as to measure the circumference, historians think was solved the same way as you would probably solve it, by actually rolling a circle on a surface and then measuring the distance for one revolution, A to A. (See the bold segment in following figure.)

The problem is to express the distance around the circle (circumference) as a number times the radius. You know the answer, from your previous math classes is, $C = \pi D$ or $2\pi r$ or 2 times 3.14 times the radius. (π is approximately 3.14.) You should actually determine this approximate formula by the above method. Try this using a wheel of a bike.)

For now, this formula will be called an assumption or postulate.

Teacher: Use the Class activity (above) to determine an approximation for Pi.

Postulate 13: The formula for the circumference of a circle is $C = 2\pi r$ or πD.

This postulate is proved in Chapter 6.

The real problem is to derive the approximate formula for the area of a circle in the form of **A = K times r²**, where K is a constant times the unit square, r². (Area is the measure of square units.) You no doubt recall that K is approximately 3.14. K was assigned the symbol "π", called Pi. (Euler, pronounced Oiler, suggested the symbol, which is the first letter in the Greek term for perimeter.) The formula is expressed as $A = \pi r^2$ since the radius is needed to draw the circle.

Note: Beckmann's book, *A HISTORY OF PI*, is especially interesting and relates the action of the Indiana Legislature in 1897. Any other book on the history of mathematics will also contain some information on Pi. (See Appendix #5)

Class Discovery Activity 3.2.2

The objective of the following activity is for the class to determine the approximate value of K in $A = K r^2$ using the figure below. (Copy the figure below on graph paper.)

Estimate the approximate number of squares (to the nearest tenth) within the circle on each row. The first 2 estimates are the author's.
The number of squares in the first row (top) is estimated at 4.2. The second row is estimated at 7.1.

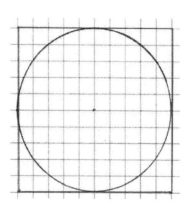

The authors estimates are:

Row 1 is 4.2 Row 2 is 7.1 Row 3 is 8.4? Row 4 is? Row 5 is?
Row 6 is? Row 7 is? Row 8 is ? Row 9 is ? Row 10 is?

Then your **Total Area** within the circle is estimated at ____Square units and now complete this TA = K (25) and solve for K. Where K is the number that you multiply the square of the radius by to get the area of the circle.
Complete this formula for K.

Your total (approximate area of circle) = K times r^2.
Total area divided by r^2 will give your calculated approximate value for Pi.
A/r^2 = K.
(Compare your value for K with the value some of your classmates have)

The correct value for K is 3.14159 to five places, but it has been calculated to over a billion places and has been proved that it is a never ending irrational number. The constant K is also defined as Pi or π as the circumference of a circle divided by the diameter, or c/d.

In Chapter 6 a more mathematical method will be given for deriving the value of Pi using geometry and your calculator. For now, we will label this formula as an assumption. You may wonder why the Postulate 14 is using 2 times the radius in place of the diameter? The answer is to draw a circle you use the radius, not the diameter.

Postulate 14: The formula for the area of a circle is A = πr^2.

This postulate is justified in chapter 6.

PLANE AND SOLID GEOMETRY ESSENTIALS

Session 3
Activity 3.3.1
Calculations

Calculate the Circumference and the Area of a circle using the following two ways. First using π as 3.14 and then using π as 3.14159. (Use your calculator Answers is given for 1a.))

1. a. A circle with a 5 ft radius.

    C   1. 31.4   2. 31.14159

    A   1. 7.85   2. 78.53975

    b. A circle with a 25 yard radius

    C   1. _____   2. _____

    A   1. _____   2. _____

    c. A circle with a 10 meter diameter

    C   1. _____   2. _____

    A   1. _____   2. _____

    d. Decision: If you were a landscape engineer, which value would you give as a workable answer to the above circles? Why?

2. The circle has a circumference of 100 ft. Solve for the radius and the area? Hint: $100 = 2(3.14) r$ and solve for r.

3. A circle has an area of 100 square ft. Solve for the radius and circumference?

4. A plan calls for a circle with a diameter of 20 feet, but your home owner in forms you to add 100 ft. to the circumference. What is the radius for the measure of the new circle?

5. The next day the homeowner (problem 4) changed his mind. The new plan calls for a circle with an additional 100 sq. ft. in the area, instead of 100 ft. in the circumference. What is the radius for the new circle?

6. What is the length of the semi-circle arc in problem 2?

7. Decision: If you had a circular walk with a 500 foot circumference and you added 50 feet to the circumference for a small flower bed area around the outside of the walk. Which of the following is your estimate as to how wide the flower bed (F) would be? Then calculate the answer using Pi as 3.14.

   a. 8 ft.   b. 10. feet   c. 11 feet   d. 6 ft.
   e. All are fairly accurate correct, but which one is the best answer? 8

8. Critical thinking Hint: Use Forms of an Implication

   Given: If I am having a good time, then I will stay at the party. Which of the following is a valid conclusion?

   a. If I am not having a good time, then I will not stay.
   b. If I do not stay, then I am not having a good time.
   c. If I stay, then I am having a good time.
   d. If I am not having a good time, then I will not stay. Answer: b

9. Decision: In the following figure: The centers of the circles are marked by the letters A to F. If each circle has a diameter of 2units, then what is the: a. perimeter of the rectangle? b. the area of the rectangle?

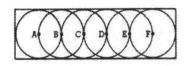

Answers to Activity 3.3.1: Applications problems, so you can check your work.

1. a. C = 31.4 ft. A = 314. Sq. ft.
       C = 31.4159 ft. A = 314.159 sq. ft.
   b. C = 157 yds A = 1962.5 sq. yds.
       C,= 157.0795 yds A = 1963.49375 sq. yds.
   c. C = 31.4 m A. 78.5 sq. m.
   d. 31.4159 m A= 78.53975 sq. m.
   e. probably 3.14159 Why?

2. a. r = 15.9 ft. to the nearest tenth. A = 793.8 sq. ft.
   b. r = 5.64 ft. C = 35.41 ft. (nearest hundredth)
   c. A = 252 r = 15.925.9 ft, (nearest tenth)

3. The answer is R = 5.65 C = 35.4 ft,
4. 15 ft.
5. 13 ft.
6. 50 ft.
7. 8 ft.
8. Then the statements are:

   a. Not GT implies not SP. (converse)
   b. Not SP implies not GT. (Contrapositive) Logical conclusion
   c. SP implies GT. (Converse)
   d. Not GT implies SP.

   (The logical valid response is "b", but all are possible since there is no statement as to what will be done if I am not having a good time.) Inverse Remember statements are true or false and conclusions are valid or invalid.)

9. a. P = 18 units   b. A = 28

PLANE AND SOLID GEOMETRY ESSENTIALS

Chapter 3 Circle
Session 4
Theorems

Understanding depends on definitions!

Unknown

First a review of circle terminology.
(T is the point of tangency.)
Use your compass or another device to draw a circle, like a coin or any round device.

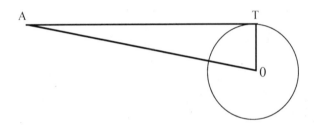

a. The name of segment AT is Tangent.
b. What is the name of segment OT? (Radius)
c. On your drawing measure AT, OT, and AO, using centimeters, to the nearest tenth. Does AT squared plus OT squared equal AO squared? (use your calculator)
d. Do you think segment OT is perpendicular to segment AT? Why? It is and the reasoning is as follows.

Recall postulate 4 stated that the shortest distance on a plane between a point and a line is the perpendicular line segment. (Check your notes or summaries if needed.)

This postulate tells us there is a shortest line segment between O and the segment AT. You think it is OT. Using indirect proof, let us assume there is another point on AT that is closer, say P, but this is false since P would have to be outside the circle by definition of a tangent. Therefore, segment PO is longer than segment OT. Hence, the only other possibility is for OT to be the shortest segment and therefore it is perpendicular.

Theorem   23: **The segment (radius) from the center of a circle to the point of tangency on the circle is perpendicular to the tangent.**

Therefore, triangle ATO is a right triangle and you automatically think

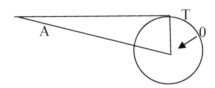

(or should) of the Pythagorean Theorem and you know that AT squared plus OT squared equals AO squared. (Pythagorean theorem) Write this in If- then form and label it Theorem 24.

Another very useful theorem in the applied world of geometry is the method of locating the center of a circle given part of a circle. (This is used by engineers and architects.)

Activity 3.4.1

Application: Say a highway engineer has three points that he wants the exit ramp to contain as the arc of a circle. Call the points A, B, C. Draw on your paper segments AB and BC and locate the midpoints. Use your ruler to locate the midpoint of segments AB and BC. From the midpoints (Label them M_1 and M_2) use your protractor and draw the perpendiculars. Where the perpendiculars intersect label it O. The point O is the center of the circle that the engineer uses to form the arc AC.

On your paper construct another case to verify the method. The question in your mind should be: "How do we know for sure that the point 0 is the center?" What must be done to prove AO equals OB and in like manner OB equals OC?

Hint: Triangles AM_1O and BM_1O are congruent also BM_2O and CM_2O are congruent, by SAS and ratio is 1. Since they are congruent then, AO equals B equals OC, therefore they are points on the circle. Hence, the point of intersection of the two perpendiculars is the center of the circle or the arc.

Theorem. 25: **The perpendicular bisectors of the chords in a circle intersect at the Center of the circle.**

Discovery Activity 3.4.2

1. Draw a circle, label the center O.

2. Draw a diameter, label it AB.

3. Draw a chord, not another diameter, with one endpoint A. Label the chord AC. Draw chord BC and measure the arcs (AC and BC) in degrees.

4. Measure the angles BAC (inscribed angle) and BOC (central angle).

5. What is your conclusion as to the measurements? Write your conclusion? The conclusion should indicate that the inscribed angle is ½ the central angle.

Definition 21: A circle's central angle is an angle with its vertex at the center of the circle.
Definition 22: The measure of the arc a central angle has the same measure as the central angle.
Definition 23: An inscribed angle is an angle where the vertex is on the circumference of a circle and the sides are chords.

Do you think your answer to #5 above is valid? Can you justify it?

Proof: Using the same figure, let BAC be any angle where AB is a diameter. Triangles AOC and BOC are what kind of triangles? Are the following true? Why?

∠AOC +∠OAC + ∠ACO equals 180

∠BOC + ∠OCB + ∠OBC equals 180

∠OAC equals ∠ACO

∠OCB equals ∠OBC

∠BOC equals ∠OCA + ∠OAC

Therefore, ∠BOC equals 2∠OAC or ∠OAC equals 1/2∠BOC.

Theorem   26: **An inscribed angle in a circle is equal to ½ the degree measure of its intercepted arc or its central angle.**

CHAPTER 3: CIRCLES
Session 5
Application- public park

Copy the following figure, and refer to it when answering the questions. In the figure, which is a plan for a park, with walks and open areas between the walks (PT and AP) and the arc. Segments are measured in **yard**s. The following information is known.

Given: Draw the figure and put the following on it.
Segment OT = 40 units, OD = 20 units, PT = 135 units
(0 is the Food Center) Draw Arc O = ACT. ΔCTB is equilateral
with AT perpendicular to BC.

In the following figure, which is a basic plan for a park with tangent PT. O is the center for arc TC. Points T, C, A, and B are on the circle. with points TCAB and TA is a diameter. The segments are the walking paths. Indicate on the figure the given information and then answer the questions.

From the given material complete the following drawing.

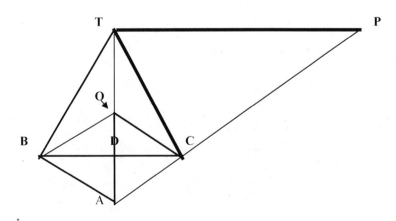

Use the drawing to answer the following:

In the above figure The line segments are the walk paths. The following is also given!

Radius OT = 40yds = OA = OB = OC, OD = 20yds, PT = 135yds, 0 is the center of the circle. **Draw the circle.**

Complete the following and be able to defend your answers!

Teacher: This could be a good home and/or class activity started in class. This activity should be worked and explained in class with the students doing the explanations and the teacher providing the hints.

Round off the measure of segments to the nearest yard and area answers to the nearest whole number of a square yard.

1. a. OP = ?   b. m∠ DBO= ?   c. DB =?   d. m∠ FTB = ?
    e. Circumference of the circle is ?   f. OP = ?
2. Arc TC = ?°
3. Arc TC = ? feet?
4. PA = ?
5. PO =?
6. 6a. m∠OCD4 = ?°   6b. m∠PAO = ?°
7. m∠TBO = ?°
8. TA = ?
9. The area of OBAC =?
10. Area of TOCP = ?
11. What is the area of the circle?
12. What is the area of the TOCPT?
13. If a person starts at O and walks to T P C and back to O, then how far was the walk?

The answers to the park problem are not given, so the teacher will have to work it and then will understand better the student's questions and provide help or even add more questions.

PLANE AND SOLID GEOMETRY ESSENTIALS

Session 6
A decision weakness!

1. Decisions are many times based on what a person sees and this can lead to contradictions. What do you see in the next picture?

W. Hill 1915 Puck Magazine

First look from the left, then from the right: What do you see?
(The High School Class Queen or the Queen's Grandmother)

2. Critical thinking: Given A implies B is valid, then which of the following are valid?

   a. A implies B and you have B, then do you have A?
   b. A implies B and you don't have A, then you don't have B?
   c. A implies B and you have A, then do you have B?
   d. A implies B and you don't have b, then you don't have A?

   Answer: #c and d are statements, be sure to understand why!

Hint: Circle A is within circle B, which reads, if in A then you are in B.

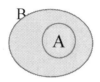

Your Summary

Terms used: Point of Tangency, chord, perpendicular bisector, inscribed angle, central angle, arc, and secant.

Theorem 20:	**The segment from the center of a circle to the point of tangency is perpendicular to the tangent.**
Theorem 21:	**The tangent from a point outside the circle squared plus the radius squared equals the square of the distance from the point to the center of the circle.**
Theorem 22:	**The perpendicular bisectors of the chords in a circle intersect at the center of the circle.**
Theorem 23:	**An inscribed angle in a circle is equal to ½ the degree measure of its central angle or ½ the degree measure of the intercepted arc.** (See Def. 11)

PLANE AND SOLID GEOMETRY ESSENTIALS

Session/Review Chapters 1 to 3

You cannot fake in mathematics, You can either prove(solve)... or you cannot.

> Jerry King
> *THE ART OF MATHEMATICS*
> From Elander's file

Review

Suggestion: Use your notes and other tools, organize your work and be able to justify your answers. If you have questions be sure to ask the teacher for explanations.

Answers are at the end of the activity for your check.

1. Equilateral triangle related questions

   a. Draw an equilateral triangle with sides measuring 2 inches using your ruler and protractor. Label the vertices A, Band C.
   b. Calculate the perimeter.
   c. Calculate the area.
   d. Add the segments for the medians to your drawing for "a." Label the medians AD, BE, and CF.
   e. Label the point of intersection in "1d" as G.
   f. What is the number of degrees in angle AGB?
   g. What is the area of triangle AGC?
   h. What is the length of segment ED?
   i. Calculate the length of segment BG?
   j. What is the physical property of point G? Hint: Center of_____.
   k. What is the measurement of angle ADB?

2. A city park is triangular in shape with sides measuring 900, 800, and 700, all in feet. The mayor asked if you could to calculate the area? Your response was yes sir. The mayor requested you have the answer by the next week, with a drawing, and the explanation as to the solution. (Suggestion: Solve this problem and then give the answer to the nearest square foot.)

   Hint: Draw the figure. Draw an altitude from one of the vertices. Label the altitude h and one segment of the base as x, then write two equations using the Pythagorean Theorem for the Area formula.

3. A park is in the shape of an isosceles triangle with the equal sides each measuring 150 feet and the equal angles are each 30 degrees. What is the length of the third side to nearest foot? Hint: Draw the figure.

4. If a square has sides measuring 10 meters, then what is the length of each diagonal?

5. The school principal needs to know the height of the school's flagpole. You are given the following facts:

   From a point 75 feet from the base of the pole two students observe the following:

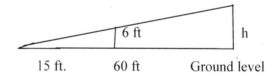

   15 ft.   60 ft   Ground level

   What is the height of the flag pole to the nearest foot?

6. Given: If you are an insurance agent, then you have studied mathematics. (This a true statement and valid)

   a. Write the converse of the given and indicate if it is true and valid.
   b. Write the inverse of the given and indicate if it is true and valid.
   c. Write the contrapositive of the given and indicate if it is true and valid.

7. A person plans a garden and has 120 feet of fencing. What is the length and width of the largest rectangular area that can be fenced? In other words, the length times the width is to be a maximum.

Suggestion: Create a table like the one below to help make your decision.
Perimeter = 2L + 2W = 120
Dimensions

Length	Width	Area(Sq. Ft)
10	50	500
20	40	800
30	?	?
?	?	?
?	?	?

Hint: 2L + 2W equals 120 or L + W = 60
Continue until you arrive at a conclusion. Write your conclusion.

8. Study your solution table in #7, then predict the maximum area if the amount of fencing were 360 feet.

9. Draw a 30-60 degree right triangle. Label the vertices A, B, and C, where C is the vertex of the right angle and angle A is 60 degrees. The hypotenuse is 10 units.

   a. Draw the altitude from C and label it CD.
   b. How many triangles are formed?
   c. A re they similar, then map the triangles and indicate the size of each angle.
   d. If the measure of the hypotenuse is 10, then what is the measure of the sides in the three triangles?

10. Why may a four-legged table wobble?

(Be sure the students understand WHY!)

Teacher: Objective of 11-13 is to point out the validity of the forms of an implication. These types are used in everyday thinking, mostly without the understanding of each. Recall: Statements are true or false and conclusions are valid or invalid. **Use the circle and the inner circle diagram for A implies B.**

Everyday thinking type problems.

Using the forms of an implication to arrive at conclusions.

Remember: If A implies B is a Valid statement and is True, then is: B implies A is valid and true? Maybe

not A implies not B is valid and true
Maybe.
If not B implies not A? Contrapositive
(Valid and true)

Now apply the above to the following and indicate which you think are True or False, or Valid or Invalid.

11. Let A be Successful and B represent not Lazy. Write each below in A implies B form and name (Statement, converse, inverse, contrapositive).

 a. Given: If you are Successful, then you are not Lazy.
Hint: A implies B or A ➜ B (Statement)

 b. If you are not successful, then you are not lazy.
 c. If you are lazy, then you are not successful.
 d. If you are not lazy, then you are successful.

12. Let F be "fun to be with" and H be "happy friends"

Statement: If you are fun to be with, then you have happy friends.
Assume this is valid. F-->H

a. Translate: If H, then F. Is it always valid? Why?
b. Translate: If not F, then not H. Is it always valid? Why?
c. Translate: If not H, then not F. Is it always valid? Why?

13. Statement: All cowboys wear handmade boots. (C-->B) Valid and true.

Given: Tex is wearing handmade boots.
Which of the following is a valid conclusion and why?

a. Tex is a cowboy   b. Tex may be a cowboy.   c. Tex is not a cowboy.

14. In a rectangle, what is the change in the area if the dimensions are doubled?
Very important concept!

Answers for review Activity (p. 101)

a. Drawing

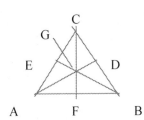

b. P = 6 inches
c. 1.73 sq. in.
d. Drawing
e. Drawing
f. 120 degrees
g. √3/3 sq. in. =.58 sq.in.
h. 1 in.
i. 2/3 √3 =1.15 in.
j. Center of gravity
k. 90 degrees

2. 268,327 or 268,328 sq. ft.
3. 259.8 or 260 ft.
    Equations: $X^2 + (H/2)^2 = 150^2$ and use Th 7.2

4. 30 feet to the nearest foot.
5. 4 ft or $(10\sqrt{2})$
6. Not always T and invalid   b. Not always True and invalid
    c. Valid and true. M implies A   d. not M implies not A and true and valid

7. 30 ft by 30 ft   Area is 8100 sq. ft.
8. 90 by 90 ft.   Area is 8100 sq. ft.
9. Draw the triangle and label.
    Angle measurement

    A = 60°   B = 30°   C = 90°   ACD° = 30°   BCD = 60°
    ADC=   BDC = 90°

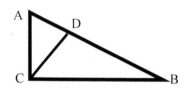

    Given: Measurements (Give answers to 2 decimal places)
                AB = 10.00   AC = 5.00   BC = 8.66   AD = 2.50
                CD = 4.33   BD = 7.50

10. Four points can determine four planes.   ABC, ABD, BCD, ACD
11. a. not valid (Converse: If B implies A)
    b. not valid (Inverse: not A implies then not B)
    c. Valid (Contrapositive: not B implies not A)
        (Remember validity does not mean the statement is true or false.)

12. a. not always (Converse)   b, not always (Inverse)   c. valid (Contrapositive)

13. Tex may be a cowboy (converse may be valid.)
    Diagram also explains it, since other people can wear hand-made (HM) boots other than cowboys.

All who wear HM Boots  cowboys

Defend your conclusion!

14. 4 times larger!

Teaching Suggestion: Carefully explain the missed problems in class or let students explain their correct method.

Understanding of the **Decision-Making problems is very important.**

Write your review of the important items
Application

Tony, a student taking Geometry, was given the following problem. His mother wanted a garden next summer and asked Tony to plan the gardens in the back yard which was 50 ft by 50 ft, a square. His father suggested the following design. Inscribe a circle in the square and have 4 areas, one in each corner of the square outside the circle. Each of the areas will have two segments of the square and an arc of a circle for the third side. **Draw the diagram for better interpretation.**

Questions:

1. That is the area of the back yard?
2. What is the area of the circle?
3. How would you locate the center of the circle?
4. What is the area of each garden? 134.58 sq. ft.
5. What is the length of the circular segment in each garden?

By the way, one garden was for flowers. one for sweet corn, one for a shade tree and one for the dog house with a shade tree.

Challenge
Write the Theorem!

Given an equilateral triangle with a side of S ft. What is the area of the circumscribed circle? (Hint: What is the radius?) Write the conclusion as a theorem! (This is probably the first time the students discovered their own theorem.) Some aids:

a. Draw the figure.
b. What is the radius in terms of S for the circle?
c. Calculate the area of the circle.
d. Write the conclusion as a theorem in if-then form.

Write your Summary

PLANE AND SOLID GEOMETRY ESSENTIALS

Chapter 4
Session 1
Volume Basics

The prior concept with applications is the area of plane figures, but we live in a three-dimensional world. This world also involves the concepts and applications of volumes. The connection is that these 3-D (three dimensional) figures consist of plane figures (2-D). For example, a box, like a shoe box, consists of six plane segments as shown in the next figure.

<table>
<tr><td>3-D View</td><td>2-D Layout View</td></tr>
</table>

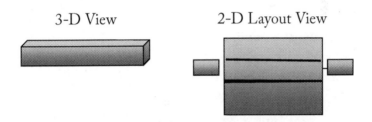

Naturally, there is a lot of mathematical thinking and reasoning in order to take a piece of flat cardboard and turn it into a box of a specified volume. What is volume?

Definition 24: Volume is the number of cubic units a 3-D figure contains.

(see cube below)

A cubic unit is a figure, like a box, where the six plane figures it contains are all squares. A square, as you recall, is a rectangle with equal sides.

Square Unit   Cubic Unit

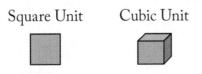

Cubic units are expressed using the most meaningful unit of measurements like cubic inches, cubic yards, cubic meters, cubic kilometers, cubic miles, or cubic hectares. What is a hectare? (See Appendix 4 or a dictionary.)

Naturally, there are mathematical formulas for calculating volumes just like there are formulas for calculating area or perimeter.

Class Discovery Activity 4.1.1

Take four sugar lumps or two pair of dice and arrange them in various shapes. No matter how you arrange the four cubes of sugar or the dice, vertically or 2 by 2 the volume is always the same.

This leads to the next assumption.

Definition 25: **Volume is an area moving through a height and the formula for volume is the area of the Base times the height, or V = Bh and the answer is in cubic units. (The measurements must be in the same units of measurement for each cubic volume.)**

Suggestion: Point out that a line may be thought of as a moving point, a plane as a moving line, and a volume as a moving plane.

What is a Tesseract? (See the interesting chapter 4 in *Mathematical Carnival* by M. Gardner, or a dictionary, or a computer search for this interesting case.)

Home Activity 4.1.2
Applications

Suggest the students make a drawing for each problem. Each of these should be explained in class, hopefully by the students!

1. A golf shoe box may measure 13 inches in length by 7 inches wide and 4 inches high, then what is the number of cubic inches?
2. A cement driveway 50 ft. long and 22 ft. wide and is 6 inches thick. What is the volume in cubic ft? How many cu. ft.? How many cubic yds?
3. If a cubic yard of cement costs $42, what is the cost of the cement for the driveway in number 2?
4. Labor cost for the driveway in number 2 is 45% of the cement cost. What is the total cost for the driveway in number 2?
5. The contractor highly recommends that a sealer be put on the cement surface after one week of drying. What is the square footage of surface in problem 2?
6. The sealer costs $19.75 per gallon and covers 450 square ft. What is the cost for the sealer needed to protect the driveway? (There is no labor cost since the owner plans to do it himself.)
7. A piece of luggage measures (inside) 3 ft. by 2ft. and 2ft high. What is the longest umbrella that can be put in the luggage to the nearest inch?

   (This is tricky!)
   Hint: The longest segment is the diagonal of the rectangular solid.
   $\sqrt{17}$ ft. or 4.1ft.

8. Draw a 3-D view of a cube and assume that each side measures 2 ft.

   a. Draw the layout view.
   b. Calculate the volume of the cube.
   c. Calculate the total outside surface area of the cube.

9. Can the surface area number of a cube ever equal the volume number? Justify your answer. **Yes**

Hint: V = A

10. Which of the following statements are true if the given is true. Given: If you are a teenager, then you are truthful.

   a. If you are truthful, then you are a teenager. (Converse)
   b. If you are not truthful, then you are not a teenager. (Contrapositive)
   c. If you are not a teenager, then you are not truthful. (Inverse)

Challenge Problem
(Ask your parents to help you!)

11. If the cube, below, is sprayed on all sides with green paint, then how many of the small cubes:

   a. Have three faces painted green?
   b. Have only two faces painted green?
   c. Have only one face painted green?
   d. Have no faces painted green?

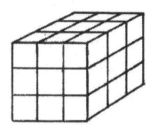

Answers for Activity 4.1.2

For your benefit and understanding. Check your answers and be prepared to justify them or your answers if different.

"The proof of the pudding is in the eating." What does that mean?

1. 364 cu. in.
2. 550 cu. ft. or 61.1 cu. Yds.
3. $855.56
4. $1240.56
5. 1100 sq. ft.
6. $48.28 or $59.25 since three gallons probably would be purchased.
7. see problem.
8. a. Drawing (6 rectangles representing the top, bottom, and 4 sides)
   b. V = 8 cu. ft. c. A = 24 sq. ft. d. Yes, when s = 6.

9. yes, when S is 6. $6S^2 = S^3$ and solve for S.
10. b. the contrapositive
11. a. 8 have three faces painted green?
    b. 12 have two faces painted green?
       6 have one face painted green?
       1 has no faces painted green?

Summary
(Add your own comments as to what you think is importance)

An important concept is to visualize and draw 2-D layouts from 3-D figures.

3-D View   2-D Layout View

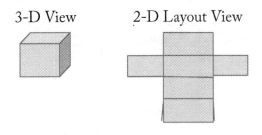

Definition 24: Volume is the number of cubic units a 3-D figure contains.

Postulate 15: Volume is a base area moving through an altitude or height.

Discovery Activity 4.1.3

Conclusions are made or should be made after collecting information from as many creditable sources as one can find. In today's world these sources are TV, magazines, books, lectures, radio reports and experts plus friendly discussions and unfortunately the unidentified social media statements.

Number Theory, it has been said, not only turns many students to math, but also results in improving their thinking ability. The following type of problem requires data organization, observation, plus discovering a conclusion and then testing the discovery.

What is the sum of the first 50 counting numbers? The following is one induction approach to the solution. (Organization and observation can help to discover an easier method.)

(N) Case	Numbers	Sum (S)
1	1	1
2	1+2 =	3
3	1+2+3 =	6
4	1+2+3+4 =	10
5	1+2+3+4+5 =	15
6	?	?

Continue in like manner until you see an easy way to arrive at the sum of the first 50 counting numbers.

The answer is $S = N(N+1)/2$ for the sum for the first N counting numbers, hence the first 50 is $50(51)/2 = 1275$.

PLANE AND SOLID GEOMETRY ESSENTIALS

Chapter 4
Session 2
Prisms & Cylinders

In life you learned to craw before you walked. In Math, you learn to understand 2-D figures before 3-D figures.

In the previous session, the following evolution was pointed out.

A moving point generates a line segment measured by its length. (One dimension)

A moving line segment can generate a plane surface measured by area.

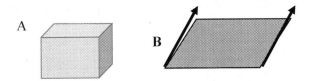

A   B

A moving line segment generates a plane. Figure B

Formulas   Can you name the figures?

$A = (½)bh$

A moving section of a geometric plane may generate a volume. Figure A (Why the condition "may"?)

A line is one-dimension in theory, where a surface is two-dimensions, and a space or volume is three-dimensions. We live in a three-dimensional world and paradoxically, we study the one and two-dimensional geometry first to really understand the three-dimensional applications. Examples of true 2-dimensional figures are shadows, or images projected on a screen. This chapter will cover more basic mathematics used in the 3-dimensional world. Architecture is a profession that is both challenging and creative. The following picture is of one of the old beautiful buildings on the campus of the University of Montana.

From the Elander's file

Take time to notice some of the types of architecture in your community. (This could be an informative research project by a group of students, which would add considerably to the student's resume'.) Can you imagine the mathematics involved in the construction of this U of MT building? (**Did you notice the M on the mountain?**) Estimating the number of bricks alone would be challenging! All the numbers and costs must be carefully calculated and a final estimate submitted in order to be awarded the contract. The study of Geometry will help you understand how the "game" of Mathematics is used in everyday situations. In previous sessions you investigated the formulas for the area of triangles, parallelograms, trapezoid, and circles. Write the formula for the area of each:

Triangle
Parallelogam
Trapeziod
Circle

Copy the formulas in your notes and be sure to name the figure the formula is associated with. The reason for the above is that the concept of volume is, basically, an area moving through a height. This means you have to calculate the area of the base and then multiply it by the height. This volume figure is identified as a prism, if the base figure is a polygon, and the figure is a cylinder if the base figure is a closed curve. But first a definition for volume is reviewed.

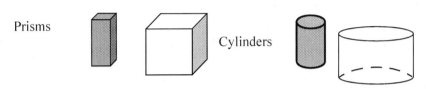

From Definition 24 the volume of a prism or cylinder is the number of cubic units it contains. The basic cubic unit of volume is one unit of length by one unit of width by one unit of height.

Example: A cube one inch by one inch by one inch is a one cubic inch.

How many cubic inches would the above cube or box contain if it were two inches on an edge? How many if it were three inches on an edge? The answers are the volume would be 8 cubic inches, and if the edge is three inches, the volume would be 27 cubic inches. Figures of this type are called prisms, if the base is a polygon, and they are called cylinders if the base is a closed curve like a circle. In this course, only special cases of prisms and cylinders will be studied.

Since there is no way to prove this apparent logical conclusion for the nth case at this stage of mathematics, this is why they were labeled postulates and since we know the formulas for some basic figures, we can write the formulas as theorems.

Postulate   15:   The volume of prism or cylinder is the product of the area of the base times the height or $V = Bh$. Where B is the area of the base and h is the height or altitude. (**The measurements must be in the same units.**)

Theorem   27: **Volume formulas for prisms and/or cylinders are:**
Prism: V = (lw)h, Cylinder: V= (πr^2)h

Where h is the perpendicular distance. (The bold line segment in the figure below.)

Example: Given the following solids:

V = (lw)h   V = (π r^2) h
Prism   Cylinder

Examples:

Base	Height	Radius (r)	Height
10x10	5	5	10

Volume: **250 cubic units**   **785 cubic units**
                       (Use π = 3.14)

Can you explain the answer for the non-right prism or cylinder?

How would you calculate the height? (Teacher may show an example.) You can easily see that to calculate volume you must be able to calculate the base area and the altitude.

Activity 4.2.1

Sketch the figure for each problem, label it, and calculate the volume.

The answers for you to compare your solutions with are provided. Ask your teacher if you have a question. The rounding off for answers is another "how to" question, and very important when calculating the cost. (Use Appendix 7 if needed.)

Base figure   Height
   1. Square: Side 3 ft.   5 ft.
   2. Rectangle: Sides 6 in. by 4 in.   10 in.
   3. Circle: Radius = 5 yds   20 yds
   4. Square: Side 7.8 meters   12.4 meters
   5. Rectangle: 3.5 ft. by 8.3 ft.   6.3 ft.
   6. Circle: Diameter = 6.9 in.   4.8 in.
   7. Convert the answer to #5 to cu. meters

Activity 4.2.1 Answers:

1. 45 cu. ft.   Sketch
2. 240 cu. in.
3. 1570.80 cu. yds.

Make your sketches

4. 7544.16 cu meters
5. 183.015 cu. ft.
6. 179.39 cu. in.
7. 5.20 cu. m.

Home Activity 4.2.2

(It will great help to understand the problem if you draw the figures and use the conversion table in Appendix 4 when needed.)

1. A water heater has a circular base with a diameter of 16 inches and a height of 5 feet. What is the volume?

2. If a cubic foot will hold 7.48 gallons, then what is the number of gallons in the water heater

3. If a cubic foot of water weighs 6.25 pounds, then what is the weight of the water in problem 1

4. How many cubic inches in a cubic foot?

5. How many cubic feet in a cubic yard?

6. If a cubic foot of gasoline holds approximately 7.48 gallons of gas and a gallon of gas weighs approximately 6.9 lbs, then what is the weight of the gas in a 20 gallon gas tank?

Challenge: Summer Camp Problem

The camp cook needs 4 quarts of water to mix with a quart of orange concentrate for the morning breakfast. The problem is that all the cook has is an 8 quart, a 3 quart and 5quart containers. The cook fills the 8 quart container and by cleverly using the 5 and 3 qt. containers is able to arrive at 4 quarts of water in the 5 or 8 quart container. How did he do it?

Containers

	Eight	Five	Three
Start 1	8	0	0
move 2	5	0	3

Now continue until you get the 4 quarts in the 8 or 5 quart Container by *a variety of transfers or steps until the last move leaves 4 qts in the 5 or 8 qt. container.*

Step N	1	4	3
	4	?	?

Now the cook has the necessary 4 quarts. (How did was it done?

Answers for Activity 4.2.2

1. a. 6.84 cu. ft
2. 51.2. gal.
3. 427.5 lbs.
4. 1728 cu in.
5. 27 cu. ft.
6. 138 lbs.

Camp. The answer for one set of moves is:

Containers		
8	5	3
8	**0**	**0**
5	0	3
5	3	0
2	3	3
2	5	1
3	5	0
3	2	3
6	2	0
6	0	2
1	5	2
1	4	3

Now there are 4 quarts in the 5 qt container and the cook has his needed 4 quarts.

The following photo of a modern building in Spokane, WA, which appears to consist of parallel planes. (As was suggested prior an informative and meaningful project is to ask a student to take a few pictures of building in their community and present the pictures to the class)

From Elander's file

The real applications are combined with not only volume, but also surface area, such as painting the outside or inside of a house. The following problems will give you more practice.

Home Activity 4.2.3

(You may need to use conversion information in Appendix 4 to help solve the problems. The approximate answers are in red but the correct units are for the student to add and answers may differ due to rounding off and the value of π that is used. Ask your teacher for the method of rounding to be used.)

1. A typical shoe box measures approximately 12 inches by 5 inches and 4 inches high.

   a. Draw a sketch of the box. (Use a ruler)
   b. What is the volume of the box in cubic inches? 240
   c. What is the area of the four sides, not including the top and bottom? 160

   d. Draw the layout. (This is called the lateral area.)

   e. What is the total lateral area? 280

2. A freight car, filled with oats, measures 33 feet by 8 feet and is 6 feet high. If a bushel of oats is 1.25 cubic feet, then how many bushels of oats are in the freight car? 1267.2

3. A student's room at college measures 10 feet by 7.5 feet by 8 feet high. The student calculates the total area for painting (include the ceiling, but not the floor) and subtracts 10% of the area for windows.

   a. What is the area in sq. ft. to be painted? 675 sq. ft.

   b. If a gallon of paint will cover 400 square feet, then how many gallons should the student buy? 2 gal.

   c. What is the room's volume? 600.

4. Fans are rated by the amount of air they move per minute. Some sizes are 450 cu ft., 650 cu, ft. and 750 cu. ft. Which fan size would you suggest the student (problem 3) purchase and why? 750 cu. ft.

5. What is the number of cubic yards of dirt excavated for a cellar, 60 feet by 30 feet and 8 feet deep? 533.3cu. yd.

6. A common term used today is an **acre-foot** of water. If an acre is 4840 sq. yards and a cubic foot of water weighs 62.45 pounds, then how many tons of water in an acre-foot? (Acre-foot is defined as an acre of water one foot deep.

7. A student asked the question: Can the total surface area (number) of a cube ever equal its volume (number)? What do you think? Justify your answer.

   Hint: Area of Cube = Vol of cube

       or $6s^2 = s^3$ and solve for s.

8. You are very thirsty. Which would you prefer? A cylinder (glass) with a 2 in. diameter and 2 inches high or one that is 1 inch in diameter and 4 inch high? Guess first and then solve. (The 2 by 2 glass.)

9. a. What is the volume of a cube, if the side or edge is 6 inches? 216 cu. in.
   b. What is the length of the diagonal on the side of the square? 8.49
   c. What is the length of the diagonal of the cube? (Draw the figure and indicate the 2 diagonals.) 10.4

10. For each of the following, draw the 3D figure and the layout figure. What is the volume and the surface area of each?

   a. A cube with square base (edge of 8). 512, Surface area = ? Volume = ?
   b. Cylinder with a radius of 8 and a height of 8. 1608, Surface area = ? Volume = ?

11. A typical hot water tank has approximately a 14 inch diameter and a height of 5 ft.

   a. Draw a sketch of the tank. (Use a ruler)
   b. What is the volume of the tank in cubic feet? 5.34
   c. What is the lateral area of the tank? (Lateral area does not include the top and bottom.)? 5.24
   d. What is the total area? 8.9
   e. How many gallons does the tank hold? 40
   f. What is the weight of the water in the tank? 333
      (A gallon of water weighs 8.35 lbs.)

12. Inductive Reasoning

   5 squared is 25   Answer; n(n+1) or 2x3 and add on the number 25.
   25 squared is 625
   35 squared is 1225.
   45 squared is ?
   55 squared is ?

Do you now see the easy way to square a number ending in 5? What is 95 squared?

13. Using your conclusions from other problems, is the following number 137,924,685 divisible by 2? by 3? by 5? or by 9? (Check your answer using a calculator.) Look for a method to tell you yes or no as divisibility.

Hint: Write 5 numbers divisible by 5.
Write five numbers divisible by 3.
Write five numbers divisible by 2.
Write five numbers divisible by 9.
(Check your answer using a calculator.)

Do you see any patterns so you can predict what numbers are divisible? Write them down!

14. Dog Problems: There are a lot of dogs in Missoula, Mt. and three dogs were reported lost. The problem is to locate the houses so the dogs can be returned. The police have the dogs and the following information. Mutt's lives at house M. Bite's owners live at house B, and Show's owners are at house S. (There are 4 possible answers so draw a 4 inch line segments and locate the houses.)

The known facts are:

a. Mutts house (n) is closer to Bites (B) than to the Fire Hydrant.
b. Bite's (B) house is between Mutt's (M) and the fire hydrant.
c. Show's (S) house is farther from Bite's (B) than Mutt's (M) is from the fire hydrant. There is more than one solution?

Let the students justify their solution.
Summary

In a previous session the following evolution was pointed out.
A moving point generates a line segment measured by its length.
(One dimension) ◄────────────►
A moving line segment can generate a plane surface measured by area.

(Two dimensions)

A moving plane segment can generate a space, measured by volume.

(Three dimensions)

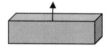

Volume is an area moving through a height.

Theorem   27: **The volume of prism or cylinder is the product of the area of the base times the height. (Height or altitude is the measurement of the perpendicular distance. The measurements must be in the same units.)**

Write your Summary.

PLANE AND SOLID GEOMETRY ESSENTIALS

Chapter 4
Session 3
Volumes of Pyramids

Many buildings are not the typical "shoe box" shape. Session 3 and 4 will explore the shapes that are pyramids or cones. The followings figures illustrate these types.

Pyramid   Cone?

The Problem and the Theory

The problem is how to calculate the volume? As you may guess, we start by calculating the area of the base figure, which you know how to do. The problem is that the cross section base area gets smaller as the base figure moves to the top or apex. This requires a different approach and is more difficult. Since we know how to calculate the volume of a prism (base area times the altitude) the prism will be used to derive a formula for the volume of pyramids. The following activity illustrates this method.

Discovery Activity 4.3.1

The following will lead to a conclusion as to the volume of a pyramid.

Teacher: **This is a great group activity! Suggest three or four students work together.** Each group of 2-4 students will need a pair of scissors and scotch tape to cut out the square and 4 triangles, so the group will have a layout to fold and form their pyramid from the sturdy sheet of construction paper or thin cardboard.

With a square, 2- inch side, and on each side of the square draw an isosceles triangle with equal sides approximately, and an altitude of 1.8 inches, as drawn below:

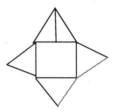

The **square** is 2 inches on a side and. each **triangle** has a base of 2 in. and an altitude of 1.7 in.

Cut out the figure on the **perimeter**!

Fold on the base of each triangle to from a pyramid and tape.

Teacher: To save time, the layouts could be done in advance! Or buy 6 identical pyramids for a class demonstration.

Your figures should be similar to the following one. Each group will need 6 of these. Now cut out the above layouts and assemble each to form a pyramid. Suggest you tape the edges to secure each pyramid.

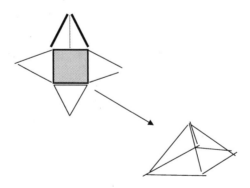

Secure each of the six pyramids using tape and assemble a cube with the 6 pyramids as shown below.

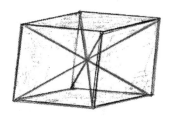

One on top and one on the botton and one on each side!

Do you see the 6 pyramids that make the cube? Keep looking! Each face or side of the cube is a base for a pyramid, therefore the volume of the six pyramids equals the volume of the cube.

We know the volume of the cube is the area of base times the altitude, but the altitude of the cube is 2h, where h is the altitude of one pyramid.

Stated algebraically: $V_p = B(2h)/6$ or $V_p = Bh/3$

Theorem   28: **The volume of a pyramid is the area of the base times the altitude divided by 3. Formula: $V_p = Bh/3$**

Theorem   29: **The lateral area of a pyramid with a square base is the surface area not including the base, which is the sum of the areas of the triangles. In the case below there are 4 triangles.**

Therefore, the Lateral Area = (1/2)4b)(h) where b is the side of the square and h is the slant height, or in our case the LA = 2bS, where S is the slant height or altitude of the triangle.

Lateral area = area of the 4 triangles

Theorem 30: **Total Area of a pyramid is the base area plus the lateral area. TA = Base area plus the area of the 4 triangles**

Most buildings are a combination of several geometric figures. The picture of the Clock Tower in Spokane illustrates this. (see Appendix 7.) How many geometric figures can you identify? The profession of Architecture requires a wide background including mathematics and liberal arts.

The key to solving these volume problems is calculating the base area. Only figures that you know the formula for will be used in these activities.

Example: If a pyramid with a square base, side 6 feet, and a height of 10 ft, then the volume is V_p = Bh/3 = 36(10)/3 or 120 cubic feet.

Home Activity 4.3.2
Volumes of Pyramids
(See Appendix 4 for Conversion ratios)

The answers to the following problems may differ due to rounding off methods, but their method may be correct. Let the students explain their procedures and defend their answers. Answers are given, but the proper units will to be added by the student.

1. Given a pyramid with a square base, side 10 inches, and an altitude of 10? Draw a sketch of the pyramid, use a ruler, and indicate altitude by a dotted line. Calculate the volume in cubic inches. 333

2. A pyramid with an equilateral triangle for a base, sides of 10 inches, and the altitude of the pyramid is 10 inches.

   a. Draw a sketch of the pyramid, use a ruler, and dotted line segments for the altitude and the hidden segments.
   b. What is the volume in cu in? 14.4

3. Pyramid A has a square base, side 10, and an altitude of 5. Pyramid B has a square base, side 5, and an altitude of 10.

   a. Which pyramid do you think has the larger volume or are the volumes equal? Record your guess.
   b. Calculate the volumes of each to check your answer to part "a." (rounded to nearest whole number.)   167   83
   c. Calculate the lateral surface area of each. LA =.200 100

4. Given: Pyramid A has a square base, side 5, and an altitude of 9. Pyramid B has a square base, side 10, and an altitude of 9.

   a. No doubt pyramid B has the larger volume, but what is your estimate for how many times larger?
   b. Calculate the volume of each and answer the question in "a."
                                                       75   300   4

5. A city park has a unique slide that is basically in the shape of a pyramid with a square base and square top. One side is for the steps to the square at the top (side of 4 ft.) of the pyramid to the edge of the base pyramid. The other three sides each have a slide. This permits a faster use for the more children.

   Three young children can be using the slide at one time. The base square has a side of 14 feet and the length of steps is 10 feet. (There is actually one like this in Missoula, MT.)

   a. Draw the 3D figure which will help to understand the problem. A top view will also help!

   b. How high is the 4 foot square above the ground? Hint: Pythagorean Theorem and similar triangles.
          (Good problem!)   8.6 ft.

   Note: The interior of the pyramid is used for storage.

6. a. How many times larger is the volume of a pyramid with a square base if the sides doubled? Guess first and then calculate.   4

   b. How many times larger is the volume of a pyramid with a square base if the altitude is tripled? Guess first and then calculate. Hint: Use a for the altitude.

                                                    3

7. The Great Pyramid of Egypt has a square base, 230 meters on a side, and height of 147 meters. Answer to the nearest whole number for the volume in cu. Yards for the pyramid (1 meter is 1.0936 yards) and the volume of your class room in cu.yds? How many classrooms would have the equivalent volume as the Great Pyramid?

Vol of GP= 3390197 cu. yds Number of class rooms?

8. A math prediction problem!

   a. What is the measure of the diagonal of a (2-dimensional) square with side s? $s\sqrt{2}$
   b. What is the diagonal of a 3-dimensional cube with a side s?
   c. Looking at the answers to "a" and "b", what would your guess for the diagonal of a 4th dimensional cube.
   (Check the dictionary or the internet for what a Tessaract is?)

Good class discussion problems!

9. Decision: A group of teenagers were discussing their pay and the amount that could earn from their summer jobs. One was paid $40 per day, another $56 per day, a third $80 per day and they were all planning to work about 30 days so they would have time for other activities. One of the group ($80 per day one) said he agreed to work for a penny the first day, 2 cents the second 4 cents the third day and so forth for just only 20 days, so he could go on a family vacation. The other teenagers thought he was

stupid! Was he? Justify your answer by calculating the amount each would earn. (Parents may enjoy this problem?)

10. Indirect reasoning: Galileo, it is said, went to the leaning tower of Pisa and dropped two objects of different weights and then claimed that objects heavier than air fall at the same rate. This, of course is not a proof, but a conclusion from a few cases. Most people at that time felt heavier objects fall faster. His proof that the objects fall at the same rate went like this. The possibilities are: the heavier objects falls faster or they fall slower, or the objects fall at the same rate. (Indirect reasoning and all the possibilities.) He than reasoned that if two objects of different weights were fastened together the result would be a contradiction.

THE CONTRADICTION: The two objects together should fall faster (The total is heavier.), but the lighter one should hold back the heavier one, since it falls slower. Hence, a contradiction, an object can't fall faster and slower at the same time. This type of reasoning leads to the same contradiction if the lighter object is assumed to falls faster.

Hence, the only possibility left is that they fall at the same rate.

(Think about his reasoning.)

He then went to the Leaning Tower of Pisa and demonstrated his conclusion!

History: A computer search for information pertaining to the Pyramids of Egypt and also for life of Galileo.

Suggestion: Several students could prepare a presentation and give a report to the class on Galileo. Extra credit.

Interesting Problem

Teacher: Record their answers.

Write the first 10 integers using 4 four 4s.

1=44/44 or (4+4)/(4+4)	(example)
2 = ? (4x4)/(4+4)	6 = ?
3 = ?	7 = ?
4 = ?	8 = ?
5 = ?	9 = ?
	10 = ?

Theorem   28: The volume of a pyramid is the area of the base times the height or altitude divided by three. The formula is $V_p = Bh/3$.

What is the formula for the volume of a cone?

Art: Draw the 3-D views and the lay outs of a pyramid, cone, cube, and cylinder showing the lateral and total area of each.

Write your definition and/or explanation of Indirect Reasoning.

GEOMETRY ESSENTIALS
FOR
PLANE AND SOLID GEOMETRY

Chapter 4
Session 4
Cones: Volumes and areas

Mathematics is like a mighty tree with numbers (counting numbers) for its roots. Arithmetic grows on numbers, algebra on arithmetic, geometry on arithmetic and algebra, analytic geometry on arithmetic, algebra, and geometry. Calculus builds on all of these. It is a tree that grows in time, fertilized by the minds of mathematicians and the applied needs of society.

unknown

Quote from from a NCTM or SSMA meeting.

Theorem 28: states that the volume of a pyramid is the area of the base times the height divided by three. The formula for the area of a circle was postulated as $A = \pi r^2$ in Chapter 3 and will be proved in Chapter 5. Since the cone is basically the same as a pyramid, but with a circular base, the formula uses the same concept.

Theorem 31: The volume of a cone is the area of the base times the height (altitude) divided by three. Formula: $Vc = (1/3)\ \pi r^2 h$. (r and h are in the same units)

In this Session only cones with circular bases will be studied.

Example: Given a cone with a circular base, radius 5 cm, and a height of 10 cm. What is the volume?

Method for solving:

Write the formula: $Vc = (1/3)Bh$ or $(1/3)\pi r^2 h$

Substitute the values for r and h: $Vc = (1/3)(3.14)(25)(10)$
$$Vc = 261.7 \text{ cu. cm.}$$

The answer, in this case, is rounded to the nearest tenth and Pi as 3.14.

Activity 4.4.1
Volumes of Cone problems

(Use your calculator and conversions tables in Appendix 4, if needed. Answers are given but without the name of the units in some cases. Answers will depend on the value used for π. and how you rounded off the numbers.) **Teacher may advise!**

1. An ice cream cone (empty) has a height of 5 inches and a base with a radius of 2 inches.

$Vc = (1/3)(3.14)(2.)^2(5)$
Complete the calculations and add proper names.

2. A pine tree is shaped like cone. If the base is 40 feet in circumference and the height is 25 feet, then what is the approximate volume of the cone shape tree? (Answer to the nearest whole number) Hint: Given the circumference, can you solve for the radius? The following picture illustrates cones in nature. (Answer to the nearest whole number.)

Project: Student showing picture of cones in the community.

The photo is from Elander's File

Answer to the nearest whole number is 1052 cu. ft.

3. An Indian tepee has a diameter of 10 feet and a height of 12 feet. What is the volume of the space inside the tepee? 942

4. A farmer had to pile some of his surplus corn on the ground. The resulting conical shape of the pile is estimated to be 15 feet in diameter and 8 feet high. The farmer asked his son, who has taken geometry, to calculate the number of bushels of corn in the pile.

    Hint: A bushel is approximately 1.24 cu. feet. Solve for the following.

    a. Number of cu. ft.   471
    b. Number of bushels.   584
    c. What is the value of the corn, if the corn at $2.31 per bushel? $1349

    Note: If you know the current price per bushel of corn, use it. $?

5. Decision Making: Which of the following is a better buy for a cold drink? A cone shaped cup that has a 3 in. diameter base and a height of 4 inches or a cone shaped cup with a 2.4 in. diameter base and height of 6 in. The 4 in. high cone filled costs $1.50 and the 6 in. high cone filled costs $1.75. Justify your decision. Which do you think is the better buy?

What is the cost per cu, in. in each case? Was your better buy selection correct?   9.42   9.04   .159   1.94

6. A cone with a circular base has a diameter of 6 feet and an altitude of 4 feet. (dotted segment)

a. What is the volume of the cone?
    37.68 or 12 Pi

b. What is the distance from a point on the circle to the apex or top of the cone? Hint: Right triangle 5

Decision: Which will give the greater volume for a cone, double the radius or double the altitude? Justify your answer. Radius (Why?)

Class Activity 4.4.2.
Surface (lateral) area and the total area of a cone

1. Draw a cone, indicate the radius and the altitude. Then draw the lay out to show the base area and the lateral area and the total area.

2. What is the lateral area for a similar cone as in number 1, if the base radius is 3 and the altitude of the cone is 4? See the figure below.

Cone
Layout for the cone.

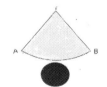

Basic one   Total surface area

Observe: The arc AB is the same length as the circumference of the base circle.

Teacher: The following may need to be explained several times!

The Formula for the total area of a cone is $A_t = \pi r^2 + \pi rS$ where S is the **slant height**, AT in the yellow figure above. The total area is the sum of the areas of the base circle and the triangular shaped lateral area, where the base of the curved triangle is the length of the arc and the height (AT) is the radius of the **arc.** (The formula for the base of lateral area (curved triangle) is $2\pi r$ which is the circumference of the base circle of the cone times the altitude. The altitude of the curved triangle is AT or the slant height S. The area of a triangle is (1/2) bh, therefore the area of the curved triangle is A= (1/2) $2\pi rS$ or $A_L = \pi rS$. (**Where S (slant height)** is the square root of ($r^2 + h^2$). Do you understand where the πrS came from?

Theorem 32: **The formula for the lateral area of a cone is $L_A = \pi rS$**

Hint: Imagine the surface area (The yellow area) of the cone divided into an **infinite number of triangles** where the total base segment is the arc of a circle. The arc of the circle is the circumference of the base circle. The area of the yellow area is ½ (h) times the base arc or $L_A = \pi rS$.

The total area is the sum of the lateral area plus the area of the base, lateral and base Total Area = $\pi rS + \pi r^2$. S is the slant height which is calculated by using the Pythagorean Theorem. The arc is the circumference of the base. Applying the formula for the area of a triangle (A = bh/2) and substituting S for the b, results in the formula $L_A = \pi rS$ and the Total Area is (1/2)$2\pi rS + \pi r^2$.

Theorem 33: **The total surface area of a cone is $\pi rS + \pi r^2$ where S is the slant height. Area = $\pi rS + \pi r^2$ sq. units.**

Draw a diagram showing the total surface area of a cone and write the formula for the total area? Total Area = $A_t = \pi r^2 + \pi rS$. What does S stand for?

Home Activity 4.4.3
(Add the proper units to your answers)

1. Native American tepee has a diameter of 10 feet and a slant height of 13 feet.

    a. What is the measure from the center of the base circle to the top of the teepee? Hint: Pythagorean Theorem 5
    b. What is the lateral area of the teepee? 1224.6
    c. What is the total area (lateral area plus the base or floor area) of the teepee? 1303.1

2. A merchant advertises two sizes of conical shaped cups for cold drinks. One is 4 inches high with a diameter of 2 inches and the other one is 4 inches high with a 3inch diameter. The drink using the small cup costs $2.00 What should the drink in the large cup cost based on the ratio of their volumes? $ 4.42 rounded to 4.40

3. Decision Making

    The number of students, who participated in the school's three Major sports, was requested by Board to report at the next meeting. The following is what the Board wanted, the number who participated in each sport. The following information is what the coach gave the superintendent, Can you help the superintendent to answer the above?

    75 students participated in A
    126 students in B
    125 students in C
    10 students participated in all three
    20 students in A and B
    50 students in B and C
    and 30 students in A and C.

    The Board wanted a more defining picture.

From this information can you derive the number who participated in only sport A, only in B, and only in C, plus the number in only 2 sports in two sports, and the number in only one sport by completing the following diagram?

Venn Diagram Problem

Hint: Draw on your paper three intersecting circles and from the given complete the numbers for each section in the diagram and then answer the questions.

Use this type of diagram.

Sport A   Sport B

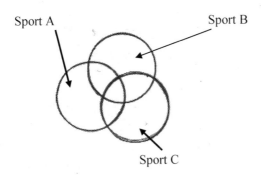

Sport C

Fill the sections indicated with the numbers. Where is the section that shows the students in three sports?

a. How many participated in only one sport, and where is it on the Venn Diagram?
b. How many participated in only 2 sports A and B? In only A and C? In only B and C?
c. How many participated in all three sports?
d. What is the total number of students out for sports?

4. Decision-Making Problem

Decisions or conclusions based on Observation are often used! Assume you are in the decorating business and are asked to estimate the cost of painting the 3 shaded areas in the design below. The paint is the same price per gallon for each color, but the cost of each color varies according to its area. What would you, from following diagram, inform the owner as to the area and cost? (The radii are determined by the dots.)

First guess the answers from observation which colors will need more paint, then by calculations.

Given: The gray circle has a radius of 3, the white circle has a radii of 4 and the blck circle a radii of 4-5.

5. a. From the figure, which colored areas do you think will need the most paint? Write down your answers, then calculate the following.

b. What is the calculated area of each color?

c. What color will cost the most for the paint?

A filling problem

6. An empty cylindrical shaped planter is 6 inches high and has a radius of 12 inches (too heavy to lift). A cone, 6 inches high with a radius of 12 inches is used to carry the water to fill the planter. How many trips would you have to make to fill the planter with water? (Guess first and then calculate the answer.)

Student Project

Take pictures of buildings in your community that illustrate the use of prisms, pyramids, cones, cylinders and other geometric figures like pine trees and create a class display. This could also provide some interesting historical facts as to the history of the community.

Activity 4.4.3 Answers

1. a. 13^2-5^2
   b. Lateral area is sq. ft. Total area is 1303 sq. ft.
2. $2.60
3. Sport A has 75 students, sport B has 126 students, sport C has 125 students, 156 students in one sport, 70 in two sports and 10 in three sports the total involved is 236.
6. Areas are: G = 9π, P = 9π, W= 7π 7.
6. Three trips

Summary

Write the definitions!
Record the Theorems

Theorem 31: The volume of a cone is 1/3 the area of the base times the height or altitude. The formula is Vc = πr^2 (h)/3. (r and h in the same units)

Student should complete the following.

Theorem 32: The lateral area is ...
Theorem 33: The total area is...

It is suggested the students make 3-D appearing drawings to illustrate the formulas.

Definitions: The meaning of lateral area and total area.

Interesting Problems
Archimedes Problem

Draw a sphere and a cone inscribed in a cylinder with a radius of R and a height of 2R. Solve for the ratios: (See cover of this text.)

a. Volume of cone is ?/? to that of the of the sphere.
b. Volume of the cone is ?/? to that of the cylinder.
c. Volume of the sphere is ?/? to that of the cylinder.

PLANE AND SOLID GEOMETRY ESSENTIALS

Chapter 4
Session 5
Volume and Area of a Sphere

TO MEASURE IS TO KNOW
Johan Kepler

Rodin's Thinker
Legion of Honor Fine Arts Museum
San Francisco, CA

It is advisable that the teacher patiently guide the students, step by step, through this session for a better student understanding.

The picture of THE THINKER was inserted since in this session the student will have to pay careful attention in order to understand.

You have come a long way from areas of parallelograms to areas of circles, from volumes of rectangular solids and pyramids to volumes of cylinders and cones. There is still one solid figure you need to investigate, the sphere. The sphere is very much a part of our culture from sports involving balls to the earth we live on. This Discovery Activity is similar to the what Archimedes used in approximately 250 B. C. or B.C. E. This should impress you as to the mathematical wisdom he developed. **Be sure to ask your questions!** You may have many as this Discovery evolves.

First, the definition of a sphere and some examples will be explained.

Definition 25: A sphere is the set of all points in 3-D space equidistant from a given point called the center.

Name a few objects that are spheres you are familiar with. Below is one very large one and one very small one.

(Geometric sphere)   (the earth)
Golf ball   North Pole

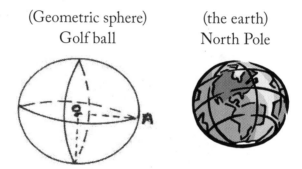

Below is a well known example and one of the author's favorites.

Compare the definition of a sphere with the definition of a circle.

A previous definition was given for a circle as the set of all points on a plane equidistant from a point called the center. What is a sphere?

Class Discovery Activity 4.5.1

Draw a sphere with a 3inch radius similar to the one below. Like the golf ball above consider the case where instead of the golf ball surface of the sphere is covered with congruent equilateral triangles and each triangle is the base of a pyramid with an altitude near the same length as the radius of the sphere. The figure below illustrates this, but with only 1 pyramid..

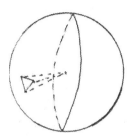

The three points of the base of the pyramid are on the surface of the sphere, the altitude of the pyramid is almost the same length of the radius of the sphere. Then the number of adjacent congruent equilateral triangles increase until the whole sphere is covered. Would you agree that the surface area of the sphere is very close to the sum of the areas of all the triangular bases of the pyramids, especially as the triangles get very small? Therefore, we could write the formula using the pyramid formula for the approximate volume of sphere as **V = (1/3) (base area) times the radius.** Think about it. **The new problem is how to find the surface (base area) area of a sphere.** Like it was stated before to solve a new problem try to reduced it to an old problem. **The surface area of hemisphere looks like the lateral area of a cone.** The formula for the lateral area of a cone is circumference times height. **For the hemisphere that would be $2\pi r$ times the height (r) or A = $2\pi r^2$, and for the area of the whole sphere [$2\pi r^2(2)$ or total area is $4\pi r^2$].** The formula as highlighted area above can now be completed by multiplying the base area times height (the radius) r. Therefore the formula for the volume of a sphere is **V = 1/3($4\pi r^3$) or V = (4/3)πr^3**

(Read the above again and ASK YOUR QUESTIONS!

So now you have the theorems for the area and the volume of a sphere.

Area of a sphere is $4\pi R^2$. (Square Units)

Volume of a sphere is $(4/3)\pi R^3$. (Cubic Units)

Theorem   34: The surface area of the sphere is $2\pi R(h)$ but h is 2R, hence the surface area of the sphere is $4\pi R^2$.

Theorem   35. The Volume of a sphere is $V = (4/3)\pi R^3$

Activity 4.5.2
Volume Problems

(Students should add the name for the units the answers are in.)

1. A 12 inch baseball means the circumference is 12 inches. In order to calculate the volume of the ball the formula requires the radius. Can the radius be determined if the circumference is known? $C = 2\pi r$ and in this case $12 = 2\pi r$. By algebra, r is $12/2\pi$ or $12/6.28$ (using pi equal to 3.14), r is 1.91 inches to two decimal places. Now use the formula for the volume and calculate the volume of a baseball. 29.2

2. If a soft ball has a 14 inch circumference, then:

    a. What is the radius? $7/\pi$
    b. What is the volume? $453.3/\pi^2 \approx 46$

3. The city of Hebron in Illinois has a spherical water tower, painted to look like a basketball. (In the early 1950s this very small school had only 16 boys and they won the all Illinois State School Basketball Tournament.) If its radius is 10 feet, then what is the volume of the water tower in cubic feet? (Use pi as 3.14) 4186.6

4. Re-work problem 3 using pi as 3.14156. What is the volume then When rounded to one decimal place? 4188.7

5. If a sphere has a radius of 10 units, then what is its volume using the ancient formula? $A = r^2 \sqrt{10}$ (Chinese about 300 B.C.)

    Calculate the volume of the same sphere using pi as 3.14. 316.2 314

6. A campgrounds has three types of cabins, (tents, rectangular cabins, and hemispheres). If the hemisphere type has a radius of 8 and an altitude of 8 feet, what is the volume? 2143.5

7. Assume the earth has a circumference of 25000 miles.

    a. What is the radius of the earth? (Use pi = 3.14) 3980
    b. What is the volume of the earth in cu. mi? 2.64x10 to the 11 power

8. If sphere A has a radius of 1 and sphere B has a radius of 2, then what is the ratio of their volumes, or what is the ratio, V_A to V_B? Guess first! 1/8

    Teacher note: Change the radii to n and m, then what is the ratio of V_A to V_B?

    Write your conclusions as a Theorem.

9. If sphere A has a volume of 1 and sphere B has a volume of 27, then what is the ratio of their radii or what is R_A/R_B? Guess first. 1/3

10. If sphere 1 has a radius of 1 and sphere 2 has radius of 2 and sphere 1 cost $2, then what should sphere 2 cost based on volume? Hint: see problem 8. Guess answer first. $8

11. A sphere has a volume of 64 cu. inches. What is the radius? (Use $\pi = 3.14$) Guess first. 3.91

12. Complete the following.

    a. One cubic foot is _____ cu. in. 1728
    b. One cubic yard is _____ cu. ft. 27

c. Assume your bed room is 10 ft x 12ft x 9 ft high, then what is the volume of your bedroom?

d. One sq. acre is 43560 sq. ft. (What is the area of a city block that you live on? An acre is about 4840 sq. yards or 43560 sq. ft.) 1080

e. One cu. meter is _____cu. yards. (1 meter is 39.37 inches or 1.09 yards.) 1.30 rounded

13. Critical thinking: Implications-Logical or illogical: 11

An interesting situation developed during the early years of the Iraq war, resulting from the use of the Flag Lapel Pin. This was a small pin depicting the American Flag that was worn on the lapel of the suit coat. Some people, many veterans were accused of being not patriotic, since they did not wear the pin. Was this a logical claim? No- why. Theorem and contrapositive relationship!

Hint: Review the forms and validity of conclusions of an If- then statement. (Theorem, Converse, Inverse, and Contrapositive)

The original statement, below, was assume valid!

If you wear the Flag Pin (A), then you are patriotic (B).

(Assume this is a valid implication.)

Write the converse of the assumption.	Not always Valid
Write the Inverse of the assumption.	Not always valid
Write the contrapositive of the assumption.	Valid

14. If you wear a green baseball cap, then you go to go to Green High School. Now write the converse, inverse, contrapositive and answer the three forms for related validity.

Activity 4.5.3
Interesting Questions

Let us assume that a rabbit runs 10 times faster than a turtle. If the turtle has a 100 yard head start, do you think the rabbit will ever catch the turtle? Let's assume the rabbit takes 10 seconds to run the 100 yds and therefore, the turtle craws 10 yd. in 10 seconds.

A limits problem

Zeno (450 BCE) argued that in order for the rabbit to catch the turtle it would have to run the 100 yds and the turtle would have completed 10 yds in the 10 seconds. After another 10 seconds the rabbit would still be a short distances behind. The race continues in like manner and therefore, Zeno argued, the rabbit will never catch the turtle.

What do you think? (Ask your parents what they think!)

Report: Do a computer search for Zeno, 495-435 B.C

Let us assume that a rabbit runs 10 times faster than a turtle craws. Then if the turtle has a 100 yd head start, do you think the rabbit will ever run the 100 yds while the turtle craws 10 yds. (fast turtle)

Rabbit -Turtle race Comment

Zeno argued that in order for the rabbit to catch the turtle it would have to run the 100 yds and the turtle would have completed 10 yds in the 10 seconds. The race continues in like manner, and therefore the rabbit would never catch the turtle, according to Zeno!

Distances: Rabbit: 100 yds, 110, 110.1
Turtle is at: 110 yds, 110.1, 110.11 Turtle is always a small distance ahead.
Considering the times and distances:
The rabbit's distance is:
R = 100 + 10 +1 +1/10 + 1/100 + 1/1000 + …. which has a limit. 11.111111111…

Turtle's distance from the rabbit's beginning point is + 1/10 +1/100 +1/1000+ 1/10000 + ... which has the same limit, and therefore the rabbit and turtle approach same limit and the rabbit will catch the turtle.

Graph the following equations. T (turtle)=100 + D and R(rabbit) = D

Therefore: Turtle's distance plus 100 = Rabbit's distance, and the limit is 100/9 or 11.11111111111....

Summary: Notes are a great aid.

Definition 25: A sphere is the set of all points equidistant from a given point, called the center.
(A circle is the set of all points on a plane equidistant from a point, called the center.)

What does Q.E.D. mean? (Some dictionaries may not have it. Keep looking!)
Quod Erat Demonstrandum

Theorem 34: The surface of a sphere is $A = 4\pi r^2$.
Theorem 35: The volume of a sphere is $(4/3)\pi r^3$ cu. Units.

PLANE AND SOLID GEOMETRY ESSENTIALS

Chapter 4
Session 6
Problems relative to a Sphere

Mathematics through the power of computers (and calculators) pervades almost every aspect of our lives.

David L. Goines

Session 5 contained the justification for the area and volume of a sphere. The theorems were labeled **Theorems** 34 and 35 and stated that the area of a sphere is $A = 4\pi r^2$ and the volume of a sphere is $4\pi r^3/3$. You certainly understand that the accuracy of the answers depends on the value you use for pi.

Comment: The quest or pursuit for the value of pi was probably the longest unsolved problem in the history of mathematics (over 4000 years) and was finally settled by Ferdinand Lindemann (1852-1939), in 1872 who proved that Pi is irrational and transcendental.

The problem still continues for the decimal approximation to the greatest number of decimal places for pi. It has been approximated to over a billion places.

The objective for this Session is to provide more information relative to spheres. This session will also emphasize that all conclusions, valid or invalid, are based on definitions, assumptions, and previously accepted conclusions or in our case theorems. It was emphasized in previous understanding.

Example: In the Preamble to the United States Constitution, it is stated:

"We the People of the United States, in Order to form a more perfect Union, establish Justice, insure domestic Tranquility, provide for the

common defense, promote the general Welfare, and secure the Blessings of Liberty to ourselves and our Posterity, do ordain and establish this Constitution for the United States of America."

What is the total number of words? T_w = ___
What is the number of words you think are undefinable? U_w = ___
What is the number of words you think that are definable? D#
What is the percent of the total that are undefinable? X%

Solve for x: $U_w = X\% T_w$

Some of the above theorems depended on the theory of limits which states that if the value of an algebraic expression approaches a value, but never exceeds (or is always less than) that value, no matter how large a value for N you try, then that value is called a limit.

The following picture is some of the remains of the Palace of Fine Arts in San Francisco. The original was built in 1915. Can you imagine the mathematics needed and the creative art for its construction. Pictures are from the authors file.

From Elander's File

A more modern structure (below) is the Space Science Museum in Oakland, Ca. Notice triangles used in the construction of the geodesic dome. Do you see the cylinders?

From Elander's File
Report suggestion: Library or computer research and report:
Archimedes (287-212 B.C.)

Questions:

How did he, Archimedes, defeat the Roman Army?
What was he wearing when he ran through the village shouting "Eureka?" Why the word "Eureka?"
What was the name of the city?
What were some of his other contributions?

Pictures from Elander's File. Look for the geometric figures. (Opportunity for a student project containing local pictures.)

From Elander's File

Can you imagine calculating the area and volume of these hot air balloons? See Appendix 7.

You know what comes next – practice in calculating problems related to a sphere.

Home Activity 4.6.1

Example: Assume a basketball has a 6 inch radius, what is surface area? Solution: $A_s = 4\pi R^2$ is the formula and the radius is 6inches. Substituting in the formula and solving:

$A = 4\pi(6)^2 = 452.16$ sq. inches.   (Using $\pi = 3.14$)

PLANE AND SOLID GEOMETRY ESSENTIALS

1. Complete the following table. Use your calculator after you write the formula for each problem and substitute the values.

Sphere	Volume	Surface area
5 in. radius	?	?
10 in. radius	?	?
20 in. radius	Predict first, then calculate.	

2. Assume a softball has a 2.54 inch diameter, calculate the surface Area and Volume?

$$Area = 20.26 \qquad Vol = 8.57$$

3. The ball in problem 3 has a leather surface that is .12 inches thick. What is the surface area of the leather ball? What is the volume of the leather?

$$A = 2.43 \text{ cu in.}$$

4. If one sphere has twice the radius of another sphere, then what is the ratio of their areas (Guess first)?   1/4

5. If one sphere has twice the area of another sphere, then what is the ratio of their radii (Guess first)?   √2/ 2

6. If the material and labor to make the cover for a ball cost $0.12 per sq. in., then what does it cost to cover a ball with a 12-inch circumference? $5.50

Challenge Program

7. Where on the earth can you travel 10 miles south, 10 miles east, 10 miles north, and be back where you started from? Which of the following is the correct answer? Parents will enjoy this also.

Possible answers: a. No place   b. North Pole   c. Equator
The answer: Many places (any point 10 miles north of the circle that has a 10 mile circumference and is north of the South pole.

8. Given this expression $(1+1/x)^x$, then what is your guess for its value when the value of x approaches a very large number?

   a. Guess first
   b. Check your guess by letting x be large values, say: 500, then x=?

(Use your calculator! Try 1000, then x=? It may not take Some large numbers.)

1,000,000, then x=?

9. The two circles below are congruent and circle B is bolted so it can't move. The objective is for circle A to rotate around B (no slippage) and be back to C again. Which of the answers is correct?

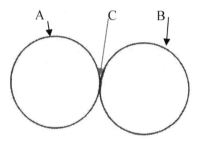

Circle A will have rotated: a. 1 b. 2 c. 3 d. 1.5 d. 2.5 revolutions.

Check your guess by experimentation. (Make a model and try it.) (let students justify their answer,)

Activity 4.6.1 Answers (M means meters)

1. a. A=314 sq. in. V = 523.3 cu. in.
   b. A = 1256 V = 4186
   c. A = 5027 V = 33.510
   h. R= 5.71 yds. V=775.3 cu.yds.
   i. R = 6M. A= 144π sq. M. V= 288 cu. M.

2. 20.26 sq. in. 8.57
3. 16.61 sq. in., 1.99 cu. in.
4. 4 to 1
5. π or the reciprocal
6. $5.50
7. Many: One place is the North Pole, but there are an infinite number of places! Another place is a point 10 miles north of the 10 mile circle of latitude from the South Pole.
8. a. 2.715568021. The value is the math constant **e**.
9. b 2.7169233932? Answers may differ due to type of calculator.
10. 2

Summary

The formulas for the surface area and volume of a sphere are:

surface area of a sphere is $4\pi\ r^2$.
the volume of a sphere is $4\pi r^3/3$.

The concept of limit.
Add your own comments to be recalled for future use.

Challenge Problems
Interesting answers!

In these problems first estimate your answer for each and then solve the problem.

a. Can the **number** for area of a square ever equal the same number as its perimeter or Area of square = the perimeter of the square?
Hint: Area = perimeter or $4S = S_2$ Now solve for S.

b. Can the volume **number** of a cube ever equal the surface area number of the cube?

Hint: $S^3 = 6S_2$ Now solve for S.

c. Can the area value of a circle ever equal the circumference value of the circle?

Complete the hint: ? = ? and solve

d. Can the volume value of a sphere ever equal the surface area value of the sphere? V = A ?

Perhaps a student or students would be interested in collecting pictures in your community of geometric applications. (Extra credit?)

Very useful 3D problem!

The author calls this the flag pole theorem and it is used by industry as well as putting up a flag pole or a basketball pole. The problem is how can you tell if the pole is perpendicular from all directions? The following justification will answer the question and will be classified as a theorem. The proof uses the theorem related to a circle and it's tangent.

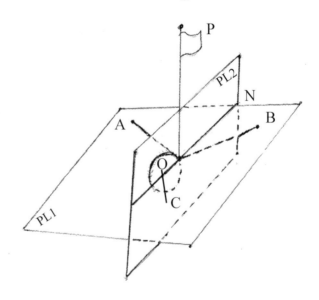

PLANE AND SOLID GEOMETRY ESSENTIALS

This theorem used by the utility companies and is very useful for everybody to know, but it should be justified. The **theorem** states that **a pole (like a flag pole) is perpendicular to the plane if it is perpendicular from only two different directions on the plane.** In the figure above the person at A, an engineer and at B is a student. Both indicate the pole looks perpendicular, but the student asks "Do we need to look from another direction to be sure it is perpendicular?" The engineer says "no", but the student asks: "Are you sure?" The engineer said, "yes, but I'll justify it for you."

Outline of proof:

Given: Plane 1 with lines AO and BO, two intersecting line segments that determine a plane, which the Pole PO is perpendicular to. Therefore:

PO is perpendicular to the Plane 1.

The justification for NO to be perpendicular to OP, another line or all lines is: Lines ON and OP determine plane P2 and OC is the radius for the circle in Pl 2 and OP is Tangent to circle with radius OC. The engineer asks the student, "What do you recall as to a tangent to a circle?" The employee says "I remember!" What did he recall?

Hence, if the pole is perpendicular from 2 directions, it is perpendicular to the plane from all directions.

Theorem 36. If a pole is perpendicular to two intersecting lines in a plane at the point of intersection, then it is perpendicular to all the lines in the plane at the point of intersection of the pole and the plane. Q.E.D. (Flag pole problem: List where it is used in your area.)

Write your summary!

Challenge

Your school has central building and two wings as illustrated below. The board voted to put the flag pole at the point equal distance from the two wings and asked your class to locate the spot.

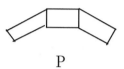

P

The board wants the pole in this area.
Where is the point and how did you locate it?
(Let you students defend their answers!)

EVERYDAY DECISION MAKING
Via
PLANE AND SOLID GEOMETRY
ESSENTIALS

Chapter 4
Session 7
Applications

Students of mathematics...the first time something new is studied, seem hopelessly confused...Then, upon returning (to the concept) after a rest, ...everything has fallen into place.

E. T. Bell
MEN OF MATHEMATICS
From Elander's File

Rodin's The Thinker
Legion of Honor Fine Arts Museum
San Francisco, CA

We live in a three-dimensional world, so to solve 3-D problems involving mathematics is a logical activity. You will encounter these types of problems sooner or later. In other words, these applications are practical. Geometry is taught not only as the basis for a review for College Entrance Exams and applications, but for the simple reason that Geometry and Critical Thinking is useful in every potential employment area that you will prepare for in the future, plus everyday decision making.

To solve the problems, use your notes, Session Summaries, calculator, and your ingenuity. If you have a question, then ask for help. Who do you ask? Anybody who you think can help you. A person can learn a lot by asking questions! It will help you to understand the problem if you sketch a figure that you think represents the problem. Organize your work so others can follow your solution method!

Activity 4.7.1
Applications Involving Spheres

Possible answers are given, but not the units of measurement, which is for the student to add.

Given: Sphere A has a radius of 2 ft. and sphere B has a radius of 6 ft.

1. What is the area of sphere A? 50.24?

2. What is the area of sphere B? 452.16?

3. What is the ratio of the areas of sphere B to sphere A? 9/1

4. What is the ratio of radii of sphere B to sphere?

5. Write a general conclusion pertaining to the ratio of the areas for two spheres with regard to the radii. r^2/R^2

6. If the ratio of the radii is 2/3, then the ratio of their surface areas is ?/?. 4/9

   Given: Sphere A has a radius of 2 units and sphere B has a radius of 8 units for problems 7-11.

7. What is the volume of sphere A? 1/16

8. What is the volume of sphere B? 2048(3.14) = 6431.

9. What is the ratio of the volumes B to A? 64/1

10. What is the ratio of the two radii of spheres B to A? 4/1

11. Write a general conclusion pertaining to the ratio of the volumes for two spheres if the ratio of their radii is x/y. $(x/y)^3$

12. Two engineers each designed a container that holds 8 cubic feet. One container was a sphere and the other a cube. Which container has the least amount of surface area? Guess first and then calculate the answers. (This type of packaging problem is used in industry.)

    Hint: Solve for the measure of the side of the cube and also the radius of the sphere, and then calculate the surface area of each. 18.4, 24

13. A tank truck consists of a cylinder and two hemispheres, one hemisphere at each end of the cylinder for aerodynamic reasons. The diameter is 8 feet and the total length of the tank is 45 feet. Use pi as 3.14.

    a. Draw the figure that represents the tank.
    b. What is radius of the hemisphere? 4
    c. What is the length of the cylinder? 37
    d. What is the volume of the cylinder? 1859
    e. What is the volume of the 2 hemispheres? 268
    f. What is the volume of the tank? 2127
    g. If a cubic foot holds 7.54 gallons, then how many gallons are in the tank? ≈ 16038
    h. Assume the tank is hauling gas and the wholesale price is $3.05 per gal, then what is the value of the gas in the tank?

    Suggestion: Use the current price of gas in your area.

Class Activity 4.7.2
More Applications and Review

Suggestion: Class activity with teacher help if necessary. Drawings will help also.

1. Archimedes favorite problem

   a. Draw a cylinder with radius R and a height of 2R.
   b. Draw a sphere with radius R.
   c. Draw a cone with radius R and height of 2R.
   d. Draw the figure showing the cone and sphere inside the cylinder. (See the text book cover.)
   e. Write the formula for the volume of each.

$$V_{sph} = \underline{\hspace{1cm}} \quad V_{cone} = \underline{\hspace{1cm}} \quad V_{cyl} = \underline{\hspace{1cm}}$$

   f. What are the ratios of the volumes?

$$V_{cyl} = ?(V_{cone}) \text{ Vcyl to } ?(V \text{ sph}) \ (V_{sph}) = ?(V_{cone})$$

Answers: 3/1 ? 2/1

   g. The volume of the cylinder is ____ times the volume of the cone.
   h. The volume of the sphere is ____times the volume of the cone.

Theorem 37: (Archimedes favorite theorem) If a sphere and a cone are inscribed in a cylinder such that the radius of the cylinder, sphere, and the cone is R and the altitude of the cylinder and cone is 2R. The ratios are 1:2:3, meaning the volume of the sphere is twice the volume of the cone and the volume of the cylinder is three times the volume of the cone.

This was Archimedes' (287-212 BC) favorite theorem and the drawing of the cylinder with the inscribed sphere and cone was on his tomb.

i. What is the volume formula for each solid figure below?

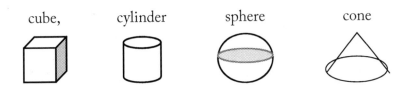

cube,   cylinder   sphere   cone

j. What are the following Volume ratios? Cyl/cyl, Cyl/Sph, Sph/Cone
k. The three figures, cylinder–sphere-cone, can all be all be inscribed in a cube with sides 2R. Draw the 3-D picture.

What is the surface area of the cube, the cylinder, the sphere and the cone?

Surface areas: Write your answers for the ratio for the surface areas as in j formula form.

Cube = 24 R^2 Cyl = 6πR^2 Sphere = 4πR^2/3
Cone = πR^2 (1+$\sqrt{5}$)   24 R^2
18.8R^2 12.6R^2 9.9R^2

2. It is reported that Lewis and Clark were very impressed with the Bull boats, the Mandam Native Americans used for transportation on the Missouri River. These boats resembled a semi-sphere covered with buffalo skins. Assume a boat is about 6 feet in diameter and a 3 feet deep (altitude). What is the volume of water displaced. Would the Bull boat hold support a 1000lbs? (The weight of the water displaced is what the boat will support.)

36π Support weight about 7257 lbs

3. a. How to become a millionaire in 31 days!

The first day you put a penny in the bank. The second day you put in 2 cents. The third day you put in 4 cents and so forth doubling the amount of the preceding day. A is the amount deposited and S is the sum of the deposits. Complete table for the 12 days. Look for a PATTERN then make predictions for the amounts on the days 28-31.

Day	Deposit	Total
1	1	1
2	2	3
3	4	7
4	8	15
5	16	31

What is the problem with this method of becoming a millionaire? Hint: The needed amounts for the last few days

b. Check your answer for the total by the following formula. This is one formula for the answer. Your pattern method may reveal an easier one!

Example on the 10th day the total will be 2 to the exponent 9 and -1 or 1023.

c. What is the sum for 31 days? (Your parents may enjoy this problem!)

4. If a rectangle has sides measuring 12 and 5, then what is the measure of the diagonals? 13

5. If a circle has diameter of 6, then what is the area of the circle? 28.3

6. If a right triangle has sides of 3, 4, 5 and the midpoints of the sides are connected, then what is the area of the internal triangle formed by the three new line segments? (Draw the figure.)

7. If the sides of a triangle are 5, 15 and x, then what is the value of x if the square root of x is an integer?

8. If two diameters of a circle (radius is 6) are perpendicular and the endpoints are connected, what is the area of the square that is formed?

9. Many times people will take a short cut across a corner of a lot or lawn. In the picture below, what is the distance saved by taking the short cut?

   Given: AB and AC are each 25 feet.

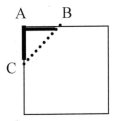

   a. What is the length of CB?
   b. What is the length of AB + AC?
   c. Should the owner be concerned?
   d. What is the distance saved in 180 school days?

10. Recall that conclusions are valid or invalid, and statements are true or false. Consider the following ad as valid:

    If it is a Q-Computer, then it is quality-built.

    a. Which of the following are **valid** conclusions and why?
       If the computer is quality-built, then it is a Q-Computer
       If the computer is not quality-built, then it is not a Q-Computer.
       If the computer is not a Q-Computer, then it is not quality-built

    b. What does the ad want you to believe is true?

Activity 4.7.2 Answers

See Activity in text for the first few answers

5. 28.3
6. 1.5
7. 16
8. 72 sq. units
9. a. 35-36 ft.   b. 50 ft.   d. 14-15 ft. per day but the lawn is ruined.
10. a.   Statement 2 is valid, but probably not true.
    b.   "If you want quality, then buy a Q-Computer." or "If you don't buy a Q-Computer, then you won't get quality."

Write your Summary

PLANE AND SOLID GEOMETRY ESSENTIALS

Chapter 5
Review
Chapters 1-4

That they (all citizens) might excel in public discussions on philosophic or scientific questions, they must be educated (rhetoric, philosophy, mathematics, and astronomy).

The Athenian Sophist School
Curriculum(480 B.C.E.)
F. Cajorie

Rodin's: The Thinker
Legion of Honor Fine Arts Museum
San Francisco, CA
From Elander's File

Review of Geometry Essentials, Chapters 1-4 with Applications

Note: The answers are provided for your benefit, but be able to justify your answers. Remember the "proof of the pudding is in the eating!"

Many of these questions involve the content related to the questions on the math portion of the college entrance exams. Some of the answers may be wrong, which may make you think! Use your notes and calculator, when needed! Make a note of the questions you miss and the related theorems.

Teacher suggestion: The missed problems may indicate a content emphasis review and an emphasis modification for next year.

Review Activity 5.1.1

1. What is the sum of the angles of a plane triangle?

2. If a theorem has been proved, then is the converse always valid? T or F

3. If the sides of a triangle are 24, 16 and x, then what does this information tell you about the length of side x?

4. a.　Do any three points determine a triangle?

　　b.　How many points are needed to determine a geometric plane?

5. If a square has a side of 10 ft then:
   a. What is the perimeter?
   b. What is the area?
   c. What is the measure of the diagonal? (Give answer to the nearest inch.)

6. If the vertex angle of an isosceles triangle is 72 degrees, then what is the measure of the other angles?

7. What is the method used for arriving at a conclusion by indirect proof?

8. If the figure is a triangle, then it is a polygon? Is the converse true?

9. Can a median of a triangle ever be outside the triangle? Explain.

10. Can an altitude of a triangle ever be outside the triangle? Explain.

11. If a ray divides an angle into two equal angles, then what is the name of the ray?

12. What is a postulate?

13. What is the test for a valid definition?

14. If two lines intersect, then the ____ angles are equal.

15. In the following figure, what is the measure of angle ACD?

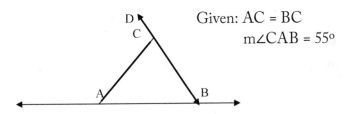

Given: AC = BC

m∠CAB = 55º

16. All logical conclusions are derived from, ____, ____, ____ and ____.

17. Similar figures in geometry have two properties, ? and ?.

18. In the following figure:

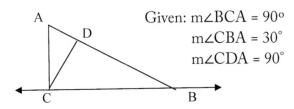

Given: m∠BCA = 90º

m∠CBA = 30°

m∠CDA = 90°

a. Map the three similar triangles.
b. If AB is 10, then calculate the measures of the other five line segments to the nearest tenth.
c. What is the ratio of the sides of the large triangle to the sides of the smallest triangle?

19. A round pizza has a diameter of 14 inches.

    a. What is the area of the pizza to the nearest sq. in.?

    b. If it is put into a square box, then what is the area of the square? Assume the pizza just fits in the box.

20. A cone and pyramid each have an altitude of 8 inches and the circumference of the base of the cone is 31.42 inches. The pyramid has a square base with a perimeter of 36 inches. What is the volume of each?

        (Answers to the nearest whole number.)

21. A large circular clock reads 4 o'clock, what is the smaller angle formed by the hands? Answer in degrees.

22. The clock in number 21 is in a 10-inch square wooden frame.

    a. What is the measure of the circle's radius?

    b. The vertex of the angle formed by the hands is at the_ of the ____

    c. The sides of the square are ____ to the circle.

    d. What is the area of the circle? (Nearest tenth)

    e. What is the area of the square?

    f. What is the area of the square outside the circle? (Nearest tenth)

    g. What is the degree measure of the smaller angle when the time is 8 pm?

23. What is the **Total lateral exterior surface** area to the nearest sq. in. of the pyramid in problem number 20?

Hint: Draw the layout.

24. If a sphere (globe) has a diameter of 20 inches, then what is the volume? (Use pi equal to 3.14)

25. What is the area of the sphere in 24?

26. A 40-inch line segment (AT) is tangent to a circle with a 30 inch radius at point T. The center of the circle is labeled O. What is the length of the segment AO?

27. In a ghost town in Montana assume your Dad found a map under a rock near the Bitterroot River (Captains Lewis and Clark and their men crossed this river on the way back.). The map read: The gold is located at the center of the circle determined by:

    a. A pine tree with "T" carved on the trunk.
    b. A rock that points North.
    c. The entrance to the cave.

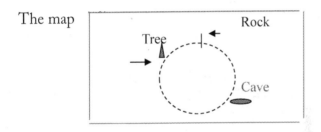

    Explain how you would find the center of the circle and the gold. Hint: see circle theorems related to bisector of a chord of a circle.

28. Draw the 3-D view of the object given the following views.

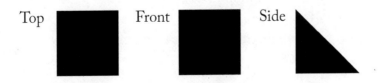

Hint: Take an object, like a box, and look at it from the three positions.

29. Is statement "b" equivalent to statement "a"? Given: Statement A is valid.

   a. If Nick does not follow directions, then he will be fired.
   b. If Nick follows directions, then he will not be fired.

30. Complete: Conclusions are ____ or not ____, but statements are ? or?

31. Complete: All conclusions are based on ____, ____, ____, and ____.

32. United States has a Republic form of government where the people elect the officials, except the President, to run the government. (Who really elects the President?) In order to have an efficient government the following are needed (among others):

   a. All voters are educated and well informed.
   b. Well informed elected representatives.
   c. News media that covers all sides of the issues.

The news media has a huge responsibility to keep the citizens informed on all sides of the issues. Informed citizens depend on that responsibility.

Statement: If you are an educated and well informed citizen, then the news media covers all sides of the issues.

Assume the above statement to be valid.
Which of the following are valid?

   a. If you have a news media that covers all sides of the issues, then you will have informed citizens.
   b. If the citizens are not informed, then the news media does not cover all sides of the issues.
   c. If the news media does not cover all sides of the issues, then the citizens will not be informed.

Answers to Review 5.1.1

1. 180 degrees
2. No
3. 8 < x < 40
4. a. No b. Three non-collinear
5. a. P = 40 ft.   b. A = 100 sq. ft. 2 in.   c. D = 10√2 or 14 ft.
6. 54 degrees
7. List all the possibilities and assume all but one. If all but one of the assumed possibilities prove false or lead to a contradiction, then the one possibility left is the correct one.
8. Yes, No.
9. No
10. Yes
11. Angle bisector
12. A statement assumed to be true is an assumption.
13. Is the definition true when reversed.
14. Opposite or vertical angles are equal.
15. 110 degrees
16. Undefined, defined, assumptions or postulates, and proven theorems or laws that follow from the first three.
17. Corresponding angles are equal in measure and corresponding sides are in equal ratios.
18. a **Map the triangles**

    b. AB = 10   CB = 8.6 or 5 √3   AC = 5   AD = 2.5   CD = 4.3   DB = 7.5
    c. Ratio is 2/1

19. a. 153.9 + sq. in.   b. 194 sq. in.
20. Vp = 216 cu. in.   Vc = 209 cu. in.
21. 120 degrees
22. a. 5 inches   b. Diagonals   c. Tangent   d. 78.54 sq. inches
    e. 100 sq. inches f. 22.46 sq. inches g. 120 degrees

23. A_T = 82.82 sq. in.

24. b. V = 4187 cu. in.

Answers may vary depending on the value of pi.

25. 1256 sq. in.
26. 50 in.
27. Draw the chords and construct the perpendicular bisectors. The intersection of the bisectors is the center.
28. Drawing.

29. No, since "b" is the inverse of "a."
30. Valid, not valid, true, false
31. Undefined terms, defined terms, assumptions and theorems Laws that follow from the first three. This is a repeat of number 16 but very important!
32. a. If you have a news media that covers all sides of the issues, then you will have informed citizens. Invalid, since you may not read or listen to the media. (Converse case)
    b. If the citizens are not informed, then the news media does not cover all sides of the issues. Invalid (The inverse case)
    c. (Valid answer) If the news media does not cover all sides of an issue, then the citizens will not be informed. Valid (Contrapositive case)

Note: The following diagram may help to understand the conclusions for a, b and c above.

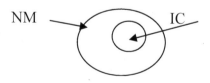

Comment: Students should keep track of exercises they missed and review them. Then repeat the exercises in a week or two. (This also emphasizes which items **need to be emphasized during the teaching.)**

Time for a fun Activity

You have worked hard and covered a lot of material so here is an interesting activity that should be first worked in class and then the students can work with their parents or friends at home. All will have fun!

The amazing Mobius Belt

Teacher: This can be a class activity where they all take part. I suggest you cut one 11 by 1inch strip (belt) for each student. The students can work in pairs, so each team with have two belts. Label each belt as below.

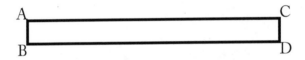

With belt number 1 tape C to A and D to B (You now have a circular belt. With a pencil or marker trace with the help of your partner a line starting at the tape and complete the complete circle. It should be evident that the belt has two sides!

With Belt number 2 twist the belt so that C to B and D to A, and tape it. Now with a pencil or marker trace a line on the belt. What do you observe? This shows that the belt has only one side.

This is called a Mobius Belt. It has only ONE SIDE! It is actually used in industry and at one time in agriculture! (Parents may be interested?)

Opportunity for a report!

Challenge Problem

If a square has 2 diagonals, then complete the following and arrive at a conclusion as to the number of diagonals for the multi-sided polygons.

Sides	diagonals	
4	2	difference
		3
5	5	
		4
6	?	
7	?	
8	?	

9   predict the number of diagonals for a 9-gon. and then check your answer by counting.

Take a break, you have worked hard!!

GEOMETRY ESSENTIALS
PLANE AND SOLID

Chapter 6
Session 1
Trigonometry Basics 1

**The advancement and the perfecting of mathematics
are closely joined to the prosperity of a nation.**

Napoleon

From Elander's File

Teacher: Students will need a scientific calculator for this session.

Euclid's book did not have a chapter on Trigonometry. (Research: Who was Euclid? Report!) Some of the terms presented in this chapter were introduced in the 16th century, but some of the ratios and the right triangle applications were known to the early Babylonians. The reason Trig is part of basic Geometry is that Trig is really an advanced way and an easier way to solve many geometric problems.

The first applications were solving for unmeasurable distances involving construction, surveying, and problems in navigation. Trigonometry resulted from the business world looking for direct and shorter routes to get their products to other countries and return with new products. As a result, the ships needed better ways of navigation. Workers needed better methods to improve construction problems. Today the problems are still construction, surveying, and navigation, but also in many others areas such as medical, electricity, atomic energy, computers, and space travel.

This is why Trigonometry with its applications in the above mentioned areas is being introduced as part of Geometry. The activities will contain examples and applications to aid your understanding and appreciation. This "playing field" today is not only on land and sea but also in space, plus its applications in a high Tech World, like in medical profession. It is more important for you to have an academic foundation in mathematics, then ever before. Keep in mind, you are just touching the tip of the mathematical "iceberg."

It was Johann Kepler (circa 1600s C.E. or A.D.), a contemporary of Isaac Newton, who said "TO MEASURE IS TO KNOW." Compare this with another quote: "TO MEASURE THE UNMEASURABLE." How can you measure the unmeasurable? From that question alone, this chapter should be interesting or at least arouse your curiosity. Can you think of a few objects that would fit in the unmeasurable list? In fact, you can't measure anything exactly. The ancient students had many similar questions just like the students today have.

This chapter begins with some basic concepts, which have their origins back in the days of the Golden Age of Greek Mathematics.

Research and Report: a. When was the "Golden Age" of Greek Mathematics? (See Boyer's A HISTORY OF MATHEMATICS or do a computer search)

                  b. The history of Trigonometry with regard to when, where, who and why.

                  c. Who was Hipparchus? Tartaglia?

Additional Interesting Reference

Teacher: Suggestion for the day before vacation: Bring in from your school library a set of interesting math books, one for each student and let the students read on Math. History and other topics for the whole period. The author has found many students checked the books out and continue to read them during vacation. (See the bibliography in the Appendix 5)

A few suggested titles:

Lieber, L: Mits, Wits and Logic
         The Education of T. C. Mits
Eves, Howard: GREAT MOMENTS IN MATHEMATICS BEFORE 1650.
Fadiman, C: The Mathematical Magpie
Kline, Morris: MATHEMATICS AND THE PHYSICAL WORLD.
(Chapter seven)
Paulos, J: I Think, therefore I laugh
Davis and Hersh Descartes Dream
Abbot: FLATAND

Or any available book on Mathematics that is in your school Library.

Teacher: If your students have had a course or an introduction to Trigonometry
         you may be able to just review parts of this session on Trig. Basics
         and work only the Activities and the theorems.

PLANE AND SOLID GEOMETRY ESSENTIALS

Chapter 6
Session 6.2.1
Trig Basics

The SINE Function

We have had several terms, which are unknown to you in just the first few paragraphs in this chapter. Can you identify the terms? The first word needing explaining or defining is the one used in the title for the chapter, Trigonometry.

What do you see in the word, "Trigonometry?" Your first answer is probably "metry", which is associated with measure. The second word you would pick out is "tri", which is associated with three as in triangle. Most students then ask where the middle part "gono" fits in. The answer is the Greeks actually used the word trigon or trigonon, which means triangular in shape. Putting these two together we have triangle measure or trigonometry. (What is etymology?) The meaning of the terms SINE and COSINE will be given at the appropriate time.

How did the Greeks develop this branch of mathematics and why? The "why" may be easier to explain then the "how." We have weather forecasters today and you probably hear their predictions every morning on the radio or TV. In the ancient days these forecasters predicted the times for planting, for harvesting, and the days for what we call holidays. We also have land surveyors, who are usually checking the boundaries for property lines or determining boundaries or key points for a freeway or Interstates. The ancient surveyors determined the boundaries after floods to determine which land belonged to the king, etc. One of their tools was a piece of rope marked off in 3-4-5 segments. Can you explain why? (Hint: Pythagorean Theorem) The surveyors of the early times were also called rope stretchers. Their tool (the 3 by 4 by 5 rope with the ends joined) when stretched at the points marked 3, 4 and 5 marks and a right triangle is formed. This was a convenient way to determine a corner for a lot. Suggest some students make this tool and demonstrate the method to the class. Today, of course, modern surveying instruments are used. (What is a cartographer?)

Discovery Activity 6.2.1

The Sine Ratio

Below is the drawing of several right triangles.

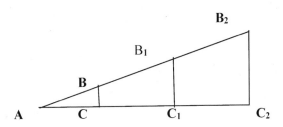

How are the triangles related? The answer is: They are similar. Why? How many triangles are there? From the fact they are similar, we also know that the corresponding sides are in equal ratio. Map the triangles and write the equal ratios. By using algebra these equations can be converted to: $BC/AB = B_1C_1/AB_1 = B_2C_2/AB_2$. This really means the following when converting the statement to if-then form. If two right triangles are similar, then the ratios of the (measures) sides opposite the acute angle divided by the (measures) hypotenuse are equal. $BC/AB = B_1C_1/AB_1 = B_2C_2/AB_2$.

Definition 26: The SINE (abbreviated SIN, but pronounce as if spelled sign) of an acute angle in a right triangle is the ratio of the length of the side opposite the angle divided by the length of the hypotenuse.

$Sin\ A = a/c$

$Sin\ B = b/c$ $=$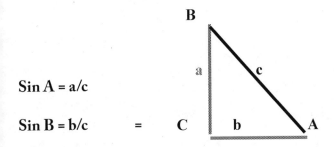

The Hindu term for this ratio was translated to SINE of the angle. The term is still used today.

Comment: The term Sine came from the Hindu with reference to the chord of a hunter's bow.

Discovery Activity 6.2.3

1. Draw a large right triangle on your paper and using your protractor draw perpendiculars so you have at least three right triangles similar to the figure below. Your large triangle should measures 6cm by 8cm by 10cm.

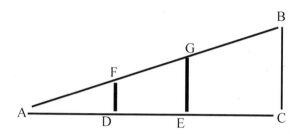

2. Complete the following table:

m<A = ? degrees (Use your protractor to measure the angle.)

Record the measure of each segment (below) using centimeters.

Answer to nearest tenth,

BC = 6cm	GE = ?	FD = ?
AC = 8cm	AE = ?	AD = ?
AB = 10cm	AG = ?	AF = ?

Evaluate the ratios. a. BC/AB, GE/AG, FD/AF
b. AC/AB GE /AC, FD/ AF
c. BC/AC, GE/AE, FD/AD.
What do you observe about the ratios?
(ratios for angles of similar triangles are equal)

Did the ratios come out the same or approximately the same? Would it make any difference if you use the metric or English system of measurement?

Fortunately, these ratios for the angles have been worked out for you, and are in your calculator and are given names. Find the sin button or key on your calculator. In order to find the sin of the acute angle your calculator must be in the degree mode. (If you don't have your book of instructions that came with your calculator, then ask your teacher to help you tell when your calculator is in degree mode.)

The definition of Sin of an angle is the ratio of the measurements of the

Steps	Display
1. Turn the calculator on	0
2. Press keys 3 then 7.	37
3. Press the sin key.	.6018

The ratio of the opposite side to the hypotenuse is 6018.

Example: How to find the sin of 37°.

The sin of an angle is the ratio of the measurements of Opposite side of the angle over the Hypotenuse and these ratios are all in your calculator! If you did not get that answer, check the book that came with your calculator and/or ask your teacher or parent. The answer to sin 37 is .60182. (To 5 decimal places)

Class Activity 6.2.4

Practice (Use your calculator)

Teacher: This is a good class activity to ensure understanding.

1. What is the sin of the following angles? (Angles are in degrees)

   a. 10   b. 13   c. 30   d. 45   e. 55   f. 60   g. 75   h. 82.
   i. What do you observe as to the value of the sin function as the angle increases?

   Answers: a. .173648 b. .224951 c. .500000 d. .707107 e. .819152
   f. .866025 g. .965926 h. .990268 i. Sin increases.

2. This time the problem will be reversed. Using your calculator, record the angles whose sines are listed below. Answer to the nearest degree.

a. Sin A = .42266   b. Sin A = .6018   c. sin A = .88295
d. sin A = .60656   e. Sin A = .9563   f. Sin A = .2079
g. Sin A = .2588   h. Sin A = .9770   i. Sin A = .35719
j. sin B.35706   k. sin B = 1.5002

Answers: a. 25°   b. 37°   c. 62°   d. 41°   e. 73°
          f. 12°   g. 15°   h. 77.7°   i. 20.92° k. no answer or error

Discovery Activity 6.2.5

The COSINE Ratio (Pronounce co-sign)

In the figure below the method used for labeling angles and sides is capital letters for vertices and small letters for sides. The small "a" represents the side opposite the vertex A, likewise the small "b" is the side opposite vertex B. This is probably nothing new from previous math courses.

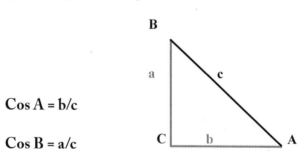

Cos A = b/c

Cos B = a/c

Definition 27: The COSINE (abbreviated COS) of an acute angle in a right triangle is the ratio of the length of the adjacent side divided by the length of the hypotenuse.

Home Activity 6.2.6

Practice

1. What is the cos value of the following angles? (The measure of the angles is in degrees)

   a. 17   b. 26   c. 30   d. 45   e. 56   f. 60   g. 72   h. 87   i. 89
   j. As the measure of the angle increases, what does the value of the cos approach?

Answers:

a. .956304   b. .898794   c. .866025   d. .707106   e. .559193.
f. .5   g. .309017   h. .052336 i. .001745   j. The value approaches 0.

2. What is the degree measure of the angle for the following?

   a. cos A = .29237   b. cos B = .48481   c. cos A = .91355
   d. cos A = .54464   e. cos B = .97437   f. cos A = .08716
   g. cos B = .70711   h. cos B = .82904   i. cos A = .35719
   j. cos B = .13917   k. cos B = 1.5002

   Answers: (in degrees rounded to the nearest degree)

   a. 73   b. 61   c. 24   d. 57   e. 13   f. 5   g. 45   h. 34   i. 69   j. 82
   k. No value or Error.

Now you are possibly beginning to think, "What are the sin or cos ratios used for?" Two examples and then some practice problems.

3. Use the given information indicated in the drawing below where side BC could be the distance across a lake, an unmeasurable distance.

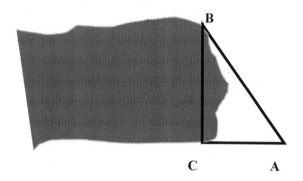

Given: m<A = 40°, BA is the hypotenuse = 110 ft.
           What is the estimated measure of side BC?

Solution: The side you need to solve for is opposite the given angle, side BC. This means you need to use the sin function, the side opposite is the side in question and the hypotenuse is known.

Write the equation, sin 40° = BC/110, then substituting the value of sin 40° from your calculator and solving for the unknown (BC) gives the value of side **a** as 70.71 feet, rounded to 2 decimal places.

.6427888 = a/110 or BC/110
Solving for "BC" using algebra:
110(.642788) = a
a = 70.706637 or 70.71 or 71 to nearest foot.

Do you think this is a good answer? Why all the decimal places? We will agree to always give the answer to fit the original data! What do you think the 110 ft. was measured with? Let's assume a measuring device measures the distance to the nearest foot, since the 110 doesn't have inches or a decimal fraction indicated. Therefore, we will give our answer to the nearest foot or 71 ft.

Another example: A ramp is 12 ft long and 5 ft high as indicated in the figure below. What is the angle of the incline (angle A)?

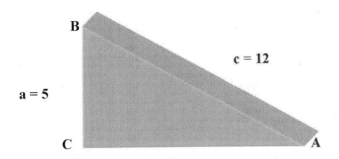

From the figure, the equation can be written: cos B = 5/12

Dividing 5 by 12 with your calculator the display reads .41667 (rounded)

Now we need to know the angle that has a cos value of .41667. Press Inverse or Shift function key on your calculator, then the cos key and the display should read 65.3756 (depending on your type of calculator) or rounded to 65.38 degrees or 65 degrees and 23 minutes (.38 of 60 is 23). For a whole number answer, the angle is closest to 65 degrees. If you solve for angle A in Sin A = 5/12, then A is equal to 24 degrees and 37 minutes. (Did you get the same values?)

In this section, you had two definitions to understand! These should be recorded in your notebook and memorized with examples.

Definition 26: The SINE (abbreviated SIN) of an acute angle in a right triangle is the ratio of the length of the side opposite the angle divided by the length of the hypotenuse.

Definition 27: The COSINE (abbreviated Cos) of an acute angle in a right triangle is the ratio of the length of the adjacent side divided by the length of the hypotenuse.

In working the following activity, keep in mind the trigonometric functions, at this point, only work with right triangles. In other words to use the sin or the cos functions you must have a right triangle. (In section 3 you will study the cases for the general angle.) If the right triangle isn't given, then you most create one in the figure by your ingenuity. Draw a figure for each problem.

Activity 6.2.7

Practice makes it easy!
Suggest this as a Class Activity

1. Using your calculator. Write the value of the sine of each of the following degrees.

Angle (degrees)	Sine	Answers
a. 5	?	.08716
b. 10	?	.17365
c. 15	?	.25882
e. 20	?	.34202
f. 25	?	.42262
g. 30	?	.5
h. 35	?	.57358
i. 40	?	.64279
j. 45	?	.70711
k. 50	?	.76604
l. 55	?	.81915
m. 60	?	.86603
n. 65	?	.90631
o. 70	?	.93970
p. 75	?	.96593
q. 80	?	.98481
r. 85	?	?
r. 90	?	?

2. Which angle or angles in problem 1 have exact values for the sine function? Answer: 30 degrees

3. What number does the value of the sin appear to approach in problem 1 as the angle gets larger? As the angle get smaller?   Answer: 1, 0

4. Using your calculator, write the value of the cos for the following angles.

Round your answers to 5 decimal places, unless the answer is exact.

Angle (degrees)

Cos		Your answer should be.
a. 5	?	.99619
b. 10	?	.98481
c. 15	?	.96593
d. 20	?	.93969
e. 25	?	.90631
f. 30	?	.86603
g. 35	?	.81915
h. 40	?	.76604
i. 45	?	.70711
j. 50	?	.64279
k. 55	?	.57358
l. 60	?	.5
m. 65	?	.42262
o. 75	?	.25882
p. 80	?	.17365
q. 85	?	.08716

Which angle or angles in problem 4 have exact values for the cos?
Answer: 60 degrees

5. What number does the value of the cos appear to approach in problem 4 as the angle gets larger? As the angle gets smaller?

Answers: 0, 1

6. Look at the answers to exercises 4 and 1, then write one additional observation.

Answer: The cos of an angle is the same as the sin of the complement.

7. Write the angles to the nearest tenth of a degree for the following.

Answers

Answers to nearest tenth	Angle
a. Sin A = .246910	14.3
b. Sin A = .376594	22.1
c. Sin A = .500000	30.0
d. Sin A = .866667	60.1
e. Sin A = .473193	28.2
f. Sin A = .654321	40.9
g. Sin A = .879605	61.6
h. Sin A = .985408	80.2

8. Read the Complete the following..

a. Cos 60 = ?
b. Cos ?   = .8660
c. Cos ?   = ?
d. Cos 70 = .3420
e. Cos 80   ?
f. Cos ?   = .6543?
g. Cos 55 = ?

Class Activity 6.2.8 (Use your calculator.)

Answers at the end of Activity for immediate check. (p. 198)

1. Convert the following to degrees and minutes.

a. 16.4 degrees is equal to 16 degrees and ? Minutes. 24
b. 61.7 degrees is equal to 61 degrees and ? minutes.
c. 57.5 degrees is equal to 57 degrees and how many Minutes?
d. 71.12 degrees is equal to 71 degrees and how many minutes?

2. Every Spring a popular past time for many young people is flying kites. If you have 500 feet of string and estimate the angle the string makes with the ground is 52 degrees, then what is the height or altitude of the kite?

    Hint: Draw and label the figure.

3. What is the measure, to the nearest degree, of the smallest angle in the famous 3-4-5 right triangle?

4. The sides of a rectangle are 12 and 20 inches.

    a. What is the length of the diagonal? (To the nearest tenth of a unit)
    b. What is the measure of the angle the diagonal makes with the side that is 20 inches? (Nearest tenth of a degree)

5. If railroad tracks are inclined at 3 degrees, then how many feet will the train be elevated in traveling one mile?

6. The sin of 30 degrees is .5. What do you think the sin of 15 degrees is? Sin of 60 degrees? Check you guesses using your calculator.

    Write your conclusion.

7. In the following triangle:

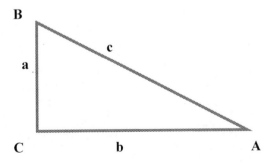

    a. What is the sin of angle A? Hint: In terms of the letters.
    b. What is the cos of angle B?
    c. What is sin A + cos B equal to?

8. If sin A = sin B, then what kind of right triangle is ABC?

9. a. Do you think the sin of 25 degrees is 1/2 the sin of 50 degrees?

   b. The sin of 25 degrees is 1/2 the sin of what angle?

10. The following figure is the side view of a ramp (not to scale) with side AB equal to 30 feet, angle A is 30 degrees and angle C is a right angle. The engineer advises perpendicular supports from E and D to side AC.

    What will be the length of the supports from E and from D to the base AC be to the nearest inch?

    The measure of BE and AD is 10 feet. Hint: Draw the figure on your paper and indicate the supports.

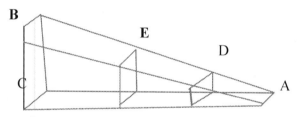

Critical Thinking: (Indirect reasoning)

11. A student divided 13 by 5 using his calculator and copied the following answer: 2.6. The student then noticed the calculator gave the following for the square root of 2: √2 is 1.4142135. The student wondered what fraction (n/d) is equal to the √2, since another calculator gave the answer 1.414213562?

    The student was really asking if the √2 can equal n/d, where n and d are integers and are in simplest form. What do you think? Simplest form means the fraction N/D can't be reduced any lower like 2/4 could be reduced to ½.

The following steps will answer the question.

a. $\sqrt{2}$ = N/D is either equal to a N/D where N and D are integers or it isn't. We will **assume** it is equal to N/D and use Indirect Proof or reasoning to answer the question.

b. Squaring each side of $\sqrt{2}$ = N/D results in 2 =N^2/D^2 which means 2 $D^2 = N^2$

c. This means N is must be even. Why?

d. If N is even, then it can be written as 2n.

e. So $2D^2 = (2N)^2$ can be written as $2D^2 = 4n^2$

f. Step e can be simplified to $D^2 = 2n^2$.

g. This means D is even.

h. Which contradicts our original assumption (see a above), and means the square root of 2 is not equal to any fraction of two integers in simplest form. (The square root of 2 is irrational.)

12. Challenge: Can (SinA)/a ever = (SinB)/b in a right triangle? Test a few cases and make your decision, then try to justify the decision for the general case. Hint; sinA = sinB or a/c = b/c which means what about the triangle?

13. $(SinA)^2 + (Cos A)^2$ =1 **Can you prove this is correct?**

Hint: Use the theorem $a^2 + b^2 = c^2$ and divide by c^2.

Activity 6.2.8

Answers

1. a. 24'   b. 42'   c. 30'   d. 7'
2. 394 ft.
3. 37 degrees
4. a. 23.3 inches   b. Angle is 31 degrees
5. 276 feet
6. Sin 15 = .2588 Sin 60 = .8660
7. a. sin A = a/c b. cos B = a/c c. sin A + cos B = 2a/c
8. Isosceles right triangle
9. a. No b. 57.7 degrees
10. 10 and 5 feet. This problem, can be done without trig and is worth doing different ways.
11. See problem
12. $(a/c)^2 + (b/c)^2 = 1$, which simplifies to $a^2 + b^2 = c^2$. $a^2 + b^2 = c^2$, Hint a/c = sinA.

Write your Summary

Definition 26: The SINE (sin) of an acute angle in a right triangle is the ratio of the length of the side opposite the angle divided by the length of the hypotenuse.

Definition 27: The COSINE (cos) of an acute angle in a right triangle is the ratio of the length of the adjacent side to the angle divided by the length of the hypotenuse.

Terms that may need defining are: Complement of an angle, Rational and irrational numbers, indirect reasoning, and degree.

PLANE AND SOLID GEOMETRY ESSENTIALS

Chapter 6
Session 2
Trigonometry
Basics 2

Hipparchus of Nicaea, (180-125 B.C.E.) compiled the first trigonometric table.

A HISTORY OF MATHEMATICS

C.B. Boyer

THE TANGENT FUNCTION

In Chapter 6 Session 2 you worked with the sin and cosine functions or ratios. The condition is that you had to have a right triangle. When you hear the phrase right triangle what important theorem comes to your mind or should? (Hint: $a^2 + b^2 = c^2$)

You are now studying the trigonometry of the right triangle. The **sin** of an angle is the ratio of the opposite side to the hypotenuse. The **cosine** of an angle is the ratio of the adjacent side to the hypotenuse. You can probably guess what the **tangent** ratio involves.

Teacher: If your students have had a course in Trig this could be a review and you could skip to Activities.

Discovery Activity 6.3.1

The Tangent of an angle

We will start just like we did for the sin and cos ratios by drawing similar right triangles. We will label the right angle C and the acute angles are labeled A and B.

Draw 3 different size right triangles with angles C, D and G all equal to 40 degrees. What is the measure of angles, B, E and H below? The **three triangles are all similar** by the angle-angle condition. This means the corresponding sides are in equal ratio. Therefore, the following ratios are equal. (side opposite to side adjacent)

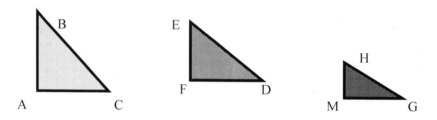

The new function is the Tangent of A (TanA) is BA/AC, TanD = EF/FD, TanG = HM/MG or the ratio of the opposite side over the adjacent side.

In each of the two lines of above statements, there are common ratios. The conclusion is that if right triangles are similar, then the ratios for the equal angles are equal. This ratio is called the tangent of the angle.

Definition 28: The Tangent of an acute angle in a right triangle is the ratio of the length of the side opposite the angle divided by the length of the side adjacent to the angle. (Tangent is abbreviated Tan.)

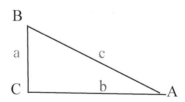

This is read as the tangent of angle A is the ratio of side a to side b.

Discovery Activity 6.3.2

Understanding the Tan. of an angle

1. Copy or draw the following three similar triangles. The triangles are 45 degree right triangles. Suggestion: BC = 3 inches, 2 inches for the equal sides.

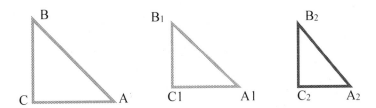

2. Measure the indicated line segments in the triangles you drew and complete the following table. Suggest you map the triangles.

	Inches	Centimeters
a. AC/BC =	?	1
b. A_1C/B_1C_1 =	?	?
c. A_2C_2/B_2C_2 =	?	?
d. What do you observe ?		

Comment: If the number displayed is not the same, check your instruction book for the calculator or ask your teacher. The calculator must be in the degree mode.

5. What is the tangent value for the following acute angles? Use your Calculator! Your answers should agree with these answers. If they don't ask your teacher.

Angel	Tangent value
5	?
15	?
25	?

35	?
45	?
55	?
65	?
75	?
85	?
90	?

6. This is the reverse of problem 5. The value of the tangent is given and you are to read the measure of the angle and from your calculator. The first problem is answered for you. Did you get the same answer?

a. Tan A = .1234, then the measure of angle A is?
   (Answer is 7 degrees) Turn calculator on and enter .1234.

   Push the "inv" or "shift or mode" key, then the "Tan" key and the display should read 7.03 rounded to the nearest hundredth. Some calculators have "Tan^{-1}" key that gives the answer directly. If you have a question, ask your teacher or read the instruction book that came with your calculator.

   (Round the answers to the nearest degree for the angle.)

   b. .1405   c. .2679   d. .4663   e. .7265   f. 1
   g. 1.80072   h. 5.1446   i. 11.4301

   Answers: b. 8° c. 15°   d. 25° e. 36° f. 45° g. 62° h. 79° i. 85°

   The symbol for degree is °, example 10° 5' 30" would be read as 10 degrees 5 minutes and 30 seconds.

An application

This example involves road construction.

In the figure below, a new road is being planned in order to move traffic more rapidly from A to B. The old route was A to C and C to B. Angle ACB is a right angle, and B cannot be seen from A. What is the angle the new road should make with line segment AC, and what is the approximate distance from A to B?

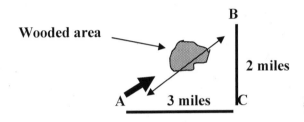

Explanation and solution:

The measure of angle CAB is needed. The known sides are opposite and adjacent with regard to angle A, therefore the tangent ratio will be used.

The equation is Tan A = 2000/3000.

The answer is 33.7 degrees or 33° and 42'. The answer is read as 33 degrees and 42 minutes (.7 of 60 is 42).

(The distance AB could also be calculated by the Pythagorean Theorem, but you need the Trig. to determine the angle for construction.)

The following questions will involve the three functions, so you will have to decide which ratio or function to solve angle CAB.

The three definitions are:

Definition 26: The SINE (SIN) of an acute angle in a right triangle is the ratio of the length of the side opposite the angle divided by the length of the hypotenuse.

Definition 27: The COSINE of an acute angle in a right triangle is the ratio of the length of the adjacent side divided by the length of the hypotenuse.

Definition 28: The Tangent (Tan) of an acute angle in a right triangle is the ratio of the length of the side opposite the angle divided by the length of the side adjacent to the angle.

Some suggestions to help solve the problems:

a. Draw the right triangle and label it with the given data.
b. Indicate on the figure the angle or side you need to find.
c. Determine which definition is required.
d. Write a trigonometric equation involving the data.
e. Solve the equation.
f. Answer the question.
g. Does the answer make sense?

Activity 6.3.3

Applications

1. Determine the Tan. value for the following angles. The angles are in degrees and you are to use your calculator. Round the answers to five decimal places.

Angles in Degrees	Tangent Value
a. 5	.08749
b. 10	.17633
c .15	.2679
d. 20	.36397

Continue for the values of Tan. for the angles of every 5 degrees increase until you reach 90 degrees.

2. What do you observe about the Tan. values as the measure of the angle increases?

   Answer: Value increases
   Question: What is the Tangent of 90 degrees? Most calculators indicate
              "Error".

3. What is the measure of an angle in degrees that has a tan equal to 1?

   Answer: 45° (Recall that ° is the symbol for degrees.

Write your summary.

PLANE AND SOLID GEOMETRY ESSENTIALS

Chapter 6
Session 3
Trigonometry for Non-Right Triangles

Teacher: This is a good problem for the teacher to lead the discussion.

Notice that the only triangles we have used in sessions 1-3 have been right triangles. **What do you do if the triangle is not a right triangle?** Naturally the mathematicians observed this and they came up with a solution that will solve all cases.

Solution 1: (Reduce the problem to a former problem.)

Step 1. Draw triangle ABC and label BD as h,

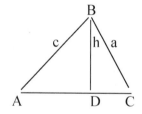

The altitude is h. You now have 2 **right** Triangles.

Step 2. Sin A = h / c
         SinC = h / a

Step 3. From step 2:   cSin A = h = aSinC
             or   cSin A = aSin C

Step 4. Multiply step 3 by a=c and you have
         (Sin A)/a = (Sin C)/ c

Do the same using the vertex B and we can conclude:
(SinA)/a = (SinC)/c = (SinB)/b

This is valid for any triangle.

Sin Theorem 38: **In any triangle:**
(Sin A)/a = (Sin C)/ c =(Sin B)/b

Activity 6.3.1

Problem using the Sin theorem

1. Given triangle ABC with AB = 10 kilometers, Angle A is 40° and angle B = 30°.

Questions:

a. Angle C = 110°   b. CB = 6.84kilometers   c. AC = 3.94 Kilometers

Rounded -off answers in red
(There will be more problems in the Activities.)
1. (sin²A) + (cos²B) = 1? Prove this for the general case.
Hint: a² + b² = c² now multiply by 1/c².Hint: for sinA substitute c/a.

The Cos Theorem

Step 1.   Draw the following figure on your Paper. Segment h is the altitude for the triangle ACB.

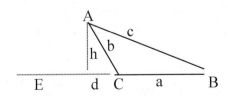

Step 2. $c^2 = (a + d)^2 + h^2$ Why?

Step 3. $c^2 = a^2 + 2ad + d^2 + h^2$

Step 4. but $b^2 = h^2 + d^2$ or $h^2 = b^2 - d^2$

Step 5. $c^2 = a^2 + 2ad + d^2 + b^2 - d^2$ **or $c^2 = a^2 + b^2 + 2ad$**

But d = bcos(angleACE)

Step 5. **$c^2 = b^2 + a^2 + 2abcosACE$**

But angle ACE is not in the original triangle!

Now complete the following Class Activity and draw the conclusion as a theorem. (Use your calculator)

Class Activity 6.3.2

A comparison as to angle ACE and angle ACB

ACE	ACB	Cos ACE. value	Cos ACB Value
10	170	.985	-985
40	140	.766	-.766
60	120	.5	-.5
80	100	.174	-.174
110	70	-.342	.342
140	40	- .766	.766

Do you see the relationship between the cos of angle ACE and Cos of angle ACB? Therefore in step 5 in the equation $c^2 = b^2 + a^2 + 2abcosACE$, the CosACE can be replaced with -Cos□BCA. Now we have an equation involving only information from the original triangle ABC. **This is the Cos Theorem, label it** Theorem #39a.

Another form of the Cos Theorem:

Theorem 39a: **The Cos Theorem:**
Form 1: Use when the side- angle- side is known)
$c^2 = b^2 + a^2 - 2abCosC$

or

Theorem 39b: **If three sides are given then this theorem solves for the angles. Cos of angle C = $(c^2 - b^2 - a^2)/(- 2ab)$.**

Suggestion: **Do these two examples in class!**
Always draw and label the figure!
Example 1: Given a = 3, b = 4 and C = 117°
What is the measure of c? (c = 5.99)
Ck the answer using the formula.
Example 2: Given a = 4, b = 3 and c = 6
What is the degree measure of angle C? (Answer: 117°)

Use the Cos Theorem **39b:**
$Cos(<ACE) = (c^2 - b^2 - a^2)/(- 2ab)$ Answer: 117°

Activity 6.3.3

Non right triangle Applications

In the following triangle the segment BC is a proposed bike trail

1. Instead of going the road route B to A and then A to C.

Given the following information: AB = 3.5 miles, A = 42° and BD is 1.5 miles. Angle A is 42 degrees and 40 minutes. What are the measurements of CB and AC?

Draw the figure and put in the perpendicular from C.

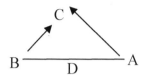

Given: A = 42° 40'
Draw the figure!

Solve for CB and AC.

2. In the following triangle, what are the answers to the following?

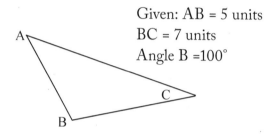

Given: AB = 5 units
BC = 7 units
Angle B =100°

Questions: a. AC between what values   a. 2< AC<12
            b. Angle A = ?   b. 48
            c. Angle C = ?   c. 32°
            d. A = ?   d. ≈ 9.3 units

3. A questionable ship is at point S and the two lookout bases are at band C. We need to know the length from B to S. We know B and C are 10 miles apart. Angle SBC is 70 degrees and angle SCB is 42 degrees. Draw the Triangle add the altitude from SD to BC. Calculate the distance from S to B.

Teacher: The students will need your suggestions on this problem. (Work it at home so you will be prepared for their questions)

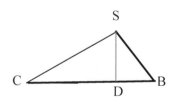

Hint: If you can solve for the length of SD, you can use the Pythagorean theorem to solve for SB and SC. Insert an altitude from S to BC and label the intersection point D on BC.

Label CD as x and BD as 10-x.

Then write what TanC is equal to and also what TanB is equal to and solve for x. Now since you have DS you can solve for SB, the distance from the lighthouse.

SB = 7.2 miles
rounded off!

Complete your summary

PLANE AND SOLID GEOMETRY ESSENTIALS

Chapter 6
Session 4
Calculating the Value of Pi

We learn the new in the light of the old.

Anonymous

Hasn't it always been a question in your mind that the approximation for pi was always given to you as 3.14 or 3.1416, but was never shown how it could be calculated? In Session 10, you were told that the ratio of the circumference of a circle to its diameter (π) was approximately 3.14, and from that page on you used approximations to work problems related to the circle. This session will provide a method to calculate the value of pi to a decimal value depending on your calculator. This will probably be a first for you! (If you know an engineer or a college math major, ask them if they ever tried to calculate π. The value of pi was calculated by Archimedes (287-212 B.C.E.) to be a value between 22/7 (3.1429) and 223/71 (3.1408).

Comment: Archimedes, some mathematicians claim, was one of the first really true mathematicians. For some of his other interesting achievements, like defeating the Roman Army, can provide an interest for an Internet, or library search. Ask your librarian for suggested references.

A recommended source is Beckmann's A HISTOTY OF Pi.

You have, no doubt, observed, that in mathematics we learn the new in the light of the old. This method for calculating the value of pi will also illustrate the meaning of that statement.

Discovery Activity 6.5.2

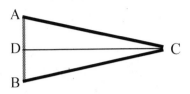

Given: Isosceles Triangle ABC with AC = 1, BC = 1, and angle ACB is 50°. CD is perpendicular to AB and is also the bisector of angle C. (Can you justify that AB is also bisected if CD bisects angle C?)

Question: Segment (AB) is what length?

Hint: Use the sin function and solve for AD.
            Sin 25 = AD/AC and by algebra AD equals = .422618261(AC), but AC is 1, therefore AD = .422618261. AB is twice AD therefore, AB is .845236522.

Caution: The above must be understood in order to work number 2!

2. Applying the method used in problem number 1 by inscribing a regular polygon ABCD in a circle and connecting the points (A, B, C, D) to the center of the circle (O) (see figure below). What is another name for segments AO, BO, CO, and DO?

    The figure below is the case when the polygon is a square. Draw the case.
a. This is really say that the value of the perimeter divided by the diameter approaches the value of pi as the number of sides in the inscribed regular polygon gets very, very large.

Using the figure below showing the regular polygon inscribed in the circle with Label M as the midpoint of segment AB.

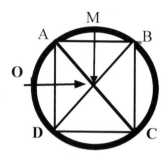

Remember the radius of the circle is 1. Now using the concept of limits, imagine the regular polygon where the number of sides is increasing and the angle AOB gets smaller and smaller and the perimeter of the polygon gets closer and closer to the circumference of the circle, the limit.

The definition for pi is the circumference of a circle divided by its diameter. Therefore, the ratio of the perimeter of the polygon of N sides as N gets very large results in the perimeter approaching the circumference of the circle and the ratio of the perimeter to the diameter approaches pi.

Now the mathematics!

a. Angle AOB is 360/n degrees and angle AOM is 180/n degrees. N is the number of sides.

b. Sin 180/n =AM/AO, AO is 1, therefore AM = 1sin 180/n. Understand?

c. The side of the polygon (AB) is 2AM and the perimeter is the product of n2AM where N is the number of sides. Understand? The perimeter, therefore is 2n(sin 180/n).

d. The ratio of the perimeter to the diameter is 2n(sin 180/n)/2 or the ratio of P/D is n(sin 180/n) and as n gets larger and larger the ratio gets closer and closer to the true value of Pi.

Example: Let n (the number of sides be 100) then the ratio is $P/D_{n=100} = 100 (\sin 1.8) = 100(.03141075908)$ or rounding the ratio value is 3.1411 (4 places) and since P/D approaches C/D which is getting closer to the correct value of Pi.

Theorem 40: One formula for Pi (π) is the limit of n(sin 180/n) as n gets very large. (N is the number of sides in the regular polygon.)

Note: **Some calculators have a Pi key for a quick approximate value of Pi.**

Class Activity 6.5.3

1. Calculate the value of PI using **Theorem** 40. Complete the following for the values for Pi. Use the formula Pi = n(sin180/n) where n is the number of sides in the regular polygon inscribed in the circle.

N values	Pi values
100	3.141075908
300	3.141535235
500	3.141571983
1000	3.141587496

These are very a good estimates for π and the formulas for the circumference and area of a circle are valid. The following were postulates which now can be called theorems, instead of postulates.

Theorem 41: C = 2πr
Theorem 42: A = πr^2

The justification follows.

The circumference of the circle is the limit of the perimeter of the polygons as the number of polygons gets very large.

C = P as n gets very large.

C = 2n(sin180/n)r, but as n gets very large n(sin180/n) approaches π, therefore C= 2πr is a theorem. QED

The Justification for A = πr² is A = (½)2(nsin180/n) r², then A = πr². Label this Theorem 40. Use the formula to calculate the following Values for pi given the number of sides for the polygon.

Number of sides	Value of pi to six decimal places
100	?
200	?
500	?

2. The figure below shows that the area of the square is between the areas of the two circles. Given: The radius of the large circle is 1.5 cm and the side of the square is 2 cm. The radius of the small circle is 1.1 cm.

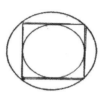

   a. What is the area of the square?
   b. What is the area of the small circle?
   c. What is the area of the large circle?

3. a. If you double the radius of a circle, how is the circumference changed?
   b. If you halve the radius of a circle, how is the area changed?

4. Some schools celebrate Pi Day, which day and month was selected for Pi Day?

5. Is .23232323... equal to a rational number? (A rational number is one that can be written as N/D, where N and D are integers.) Hint let N equal the rational number.

Rational number N = .23232323... then
100N = 23.232323... Now subtract which results in 99N =23 and therefore
N = 23/99 (Check this using your calculator.)

6. What rational numbers are the following equal to? (Use the method in #3)

.111....
.222....
.333...
.444....

Can you predict what rational number .999... is equal to?

Comments: 1. The value of π is greater than 3.1415 and less than 3.1416 to
four decimal places and is closer to the true value of pi as the
number of sides of the regular polygon increases. It took over
4000 years to completely solve the Pi problem.

**Note: F. Lindemann, about 1870, solved the questions as to Pi by proving
Pi is a transcendental number. (A transcendental is irrational and
man has no control over it.)**

Research: *HE HISTORY OF PI by* Beckmann, P.
**Especially interesting is the reference to the State of Indiana's
Legislature. Why?**

Computer search: Do a computer search for the latest value of Pi.

Activity 6.5.3

Answer

1. See problem.
2. 3.8 <4< 7.1
3. a. 4 times b. ¼ times
4. March 14th
5. 23/99
6. a. 1/9 b. 2/9 c.3/9 d. 4/9 e. 8/9

Summary

Pi is the ratio equal to the circumference of a circle divided by the diameter. Symbol is π!

A formula for Pi is the limit of n(sin 180/n) as n gets very large. (N is the number of sides in the regular polygon.

The formula for the circumference of a circle is $C = 2\pi r$ or $C = \pi d$
The formula for the area of a circle is $A = \pi r^2$.
Write and explain each of these theorems
Sin **Theorem 38**
Cos **Theorems** 39 and 40
Write the 2 forms of the Cos theorem

Theorem 43: Trig form of the Pythagorean Theorem:
$$(Sin\ A)^2 + (Cos\ B)^2 = 1$$

Here is how this was arrived at $sinA = a/c$.

1. Given: $a^2 + b^2 = c^2$
2. Divide step 1 by c^2 and write your answer
3. $(a/c)^2 + (b/c)^2 = 1$ Convert this to the Trig form answer.

Hint: In a right triangle, what is a/c equal to where c is the hypotenuse? Q.E.D. (Do you recall what Q.E.D. Means?)

Theorem 40: **A formula for Pi is the limit of n(sin180/n) as n gets very large.**
(N is the number of sides in the polygon.

Theorem 41. $C = 2\pi r$

Theorem 42. $A = \pi^2$

Challenges

1. Can the number for the circumference of circle ever be the same number as the area of the same circle? If you think it can be, then what is the radius?

2. Can the number for the area of a sphere ever be the same number for the volume of a sphere?

3. Can the number for the volume of a cube ever be the same for the area of a cube? If so, what is the measure of edge of the cube?

PLANE AND SOLID GEOMETRY ESSENTIALS

Chapter 7
Session 1
Decision-Making Skills
and the
Weaknesses

I THINK THEREFORE I AM
Rene Descartes

Rodin's "The Thinker"
Legion of Honor Fine Arts Museum
San Francisco, CA
From Elander's File

Logic-Surveys-Polls-Conclusions

Pythagoras, the teacher, paid his student three oboli (a coin) for each lesson he attended and noticed that as the weeks passed the boy's initial reluctance to learn was transformed into enthusiasm for knowledge. To test his pupil, Pythagoras pretended that he could no longer afford to pay the student and that the lessons would have to stop, at which point the boy offered to pay for his education...

Simon Singh - *Fermat's Enigma*

This chapter will help you understand why the course in geometry was established by the Greeks in Athens ($\approx$ 600 B.C.E.) when they formed the first democracy. Their objective was to help the future leaders by teaching the basics of decision making. They understood that a democracy must also have educated voters to function in a valid manner. Our "founding fathers" did also! This session will illustrate how people can become better decision makers and will help you make better decisions or conclusions as a result from applying what you have learned in Mathematics. Sound intriguing?

A branch of mathematics called LOGIC is very useful in decision-making. You will always be making decisions in the future, some correct and probably some wrong ones. This Session will just touch on a small bit of this fascinating and very practical topic.

We will begin by investigating statements in a special form, remember statements are true or false and conclusions are valid of invalid.

Discovery Activity 7.1.1

IMPLICATIONS

An implication is a statement in the form of If - then. A few examples are:

If a triangle has only two equal sides, then it is labeled as Isosceles.
If a geometric theorem is important, then it should be in the geometry books.
If the weather forecast is rain, then you should take rain gear to school.
If students do their homework, then the students will pass the course.
If you don't abide by the law, then you will be fined.
If voters are informed on issues, then beneficial laws will be passed.

The base or root word for ***implication*** is "imply" which provides the meaning for the above examples. This can be written as A implies B, where A represents the "If" part and B represents the "then" part.

This is, as you no doubt noticed, a type of short hand or an abbreviated way of writing implications. In other words, A---> B will be read "A implies B" or "If A then B." Which translation do you prefer? (Most students prefer the second one.) This type of sentence is identified as a conditional sentence or statement. You had this in previous sessions and possibly in other courses where you identified the parts of the A --->B as A, the hypothesis, and B as the conclusion. The important implications, proved in geometry are called theorems. In every day experiences they are called conclusions or even laws.

Comment: The United States Declaration of Independence by Thomas Jefferson is a great example of A implies B. (Have you read it? Read the first few pages with your parents.)

The question now is: How are implications used and evaluated? For this we will use a diagram, which will show the validity or possibly the invalidity of an implication. You will have to decide the truth or falsity of the statements. "If A, then B." can be depicted in diagram form as shown below, then the interpretation is much easier. **Remember statements are true or false, but conclusions are valid or invalid.**

A ⟶ **B means if in**

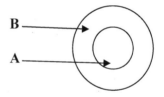

The diagram shows that if you are in circle A, then you are in circle B or A--->B. Theorems take this form. Do you recall from your previous activities, the other three forms of an implication theorem and their names?

Name	Symbolic
Converse: B---> A is read as:	If B then, A.
Inverse: ~A --->~B is read as:	If not A, then not B. The negation symbol is ~.
Contrapositive: ~B--->~A is read as:	If not B, then not A.

Using the circle diagram below for "If A, then B" and classify each of the above forms of the implication as valid or may not be valid. This diagram shows that if you are in A, then you are in B

If A then B.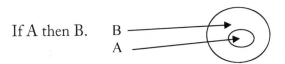

Implication

Assume that **A--->B** is true and valid. (This is the original statement.)

This means that if in circle A, then in circle B also.
Valid

Is it valid that ~B---> ~A? (**This is the Contrapositive.**) Valid
This means that if not in B then not in A.

Is it valid that ~A --->~ B? (**This is the Inverse.**) ?
This means that if not in A, then not in B. (may not be valid)

Is it valid that B --->A? (**This is the Converse,**) (may not be valid)
This means that if in B, then in A.

Do your conclusions agree with the following? Be sure to understand why, if your answers differ. **These are important relationships and are summarized below! Conclusions are valid or not valid, and statements are true or false.**

Statement	Valid?
1. Converse	May not be valid
2. Inverse	May not be valid
3. **Contrapositive**	**Valid**

Can you draw the diagram showing the case that all four are true and valid? Hint: Explain the case where A and B are concentric and identical circles

Class Activity 7.1.2

Understanding Implications

Each implication is written in one of the forms. Write the other three forms in the if-then form, and classify each as valid or possibly invalid. (Remember: Statements are true or false and conclusions are valid or not valid)

(Explain your answers to the following as to "T" or "F" cases and also indicate the validity? for the following.)

1. Statement: **If a person is a teacher, then that person enjoys students.**

   (Assume this as True and valid)
   The first three forms are answered for you.
   Write the Converse.

   Answer: If a person enjoys students, then the person is a teacher. (Could be true or false since non-teachers may also enjoy students. Conclusion: not always valid)

   Write the Inverse.

   Answer: If a person is a not a teacher, then the person doesn't enjoy students. (Could be true or false –why? Conclusion: not always valid)

   Write the Contrapositive.

   Answer: If a person does not enjoy students, then that person is not a teacher. True and conclusion is valid.

2. If the figure is a triangle, then it is a polygon. (True and valid) Write the Converse.

   Answer: If the figure is a polygon, then it is a triangle.
   Could be False and invalid

Write the Inverse.

Answer: If the figure is not a triangle, then it is not a polygon.
Could be False and invalid.

Write the Contrapositive.

Answer: If the figure is not a polygon, then it is not a triangle.
True and valid

3. Statement: **If you own real estate, then you pay taxes.**

(Assume True and Valid)
(Explain your answers.)

Write the Converse.

Answer: If you pay taxes, then you own real estate. Could be False and
not always valid.
(You don't own property, but you may still pay income tax)

Write the Inverse.

Answer: If you do not own real estate, then you do not pay taxes.
Could be False and invalid. Why?

Write the Contrapositive.

Answer: If you do not pay taxes, then you do not own real estate. True
and valid

4. Statement: **If you live in the Montana, then you live in the United States
of America.** (True)

(Explain your answers.)
Write the Converse.

Answer: If you live in the U. S. of A, then you live in Montana.
Not necessarily valid or true.

Write the Inverse.

Answer: If you don't live in the MT, then you don't live in the U.S. of A.
Not necessarily valid or true.

Write the Contrapositive.

Answer: If you don't live in the U. S. of A, then you don't live in Montana.
Valid and True

5. Statement: **If you display the flag on the 4th of July, then you are a patriot.**
(True and valid)

(Explain your answers.)
Write the Converse.

Answer: If you are a patriot, then you display the flag on the 4th of July.
Not necessarily true and may be invalid.

Write the Inverse.

Answer: If you don't display the flag on the 4th of July, you are not a
patriot. Not necessarily true and may be invalid.

Write the Contrapositive.

Answer: If you are not a patriot, then you don't display the flag on the 4th
of July. True and valid

6. Statement: **If you wear the Flag Pin, then you are patriotic.** (Consider valid and true.) It will help you to understand this situation by using the diagram below.

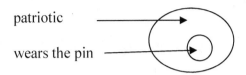

patriotic

wears the pin

Write the converse, inverse, and contrapositive and classify each as valid or not valid.

(This statement, by some groups in the 2004 Presidential election actually interpreted, the inverse as true and valid and some persons were falsely accused. **What is the inverse of the statement in #6?**

7. Can you draw a curve using straight line segments? On your paper connect 2 to 8, 3 to 7 and so forth.

$$2\ 3\ 4\ 5\ 6\quad 7\quad 8$$

8

7

6

5

4

3

2

PLANE AND SOLID GEOMETRY ESSENTIALS

Chapter 7
Session 2
The Use amd Misuse of Statistics

Statistical thinking will one day be as necessary for efficient citizenship as the ability to read and write.

H. G. Wells
From Elander's File

Rodin's: The Thinker
Legion Of Honor Fine Arts Museum
San Francisco, CA
From Elander's File

Discovery Activity 7.2.1

The Meanings of the Three Averages
Mean, Mode, and Median

The calculating of averages is fairly simple, but interpreting the results accurately can be confusing and is often misleading. You may come across numbers representing averages in the news every day, and as an informed citizen you should be able to understand the correct interpretation of these. You have probably heard the statement: "Figures don't lie, but liar's figure." It is important for you to understand the use and misuse of these averages and their implications. This session will help you interpret statements involving statistics and other information that is needed to see the true picture.

The reporting of scores or sets of data to convey the results or by a "picture" using averages and drawing conclusions from them is often used. Some examples of these are:

1. The meaning of math scores for a certain grade or class in your school.
2. The average income per family.
3. The average score per game for your school's basketball team.
4. The average stopping distance for a certain make of car.
5. The average pay per employee.
6. The average for your school's ACT or SAT scores.
7. The average property in in a city.

Let's look at the arithmetic average of the following set of scores on a high school math team exam. The team consisted of the five students.

Their scores: 100, 99, 80, 76, 70

The superintendent proudly reported to the newspaper that the average score for the 2019 team math team is 85. Does this one number give a true picture of the team's scores? Notice, no student even scored an 85?

Definition 29: The MEAN or arithmetic average for a set of scores is the sum of the scores divided by n, the number of items in the set.
Formula: Mean = (sum of n scores)/n
Ave Score = 425/5 = 85

Calculate the mean for the above math teams set of scores. Was the report correct?

The 2018 and 2019 math team scores are listed below. The superintendent reported the teams had the same average in each year.

2019 scores: 100, 99, 80, 76, 70
2018 scores: 100, 95, 91, 80, 59

Calculate the mean score for each year.

In both cases only the average (85) was reported to the parents. Was the report correct? Do you think the two teams are equal with regard to their scores? Which team would you say is better overall? Why?

Should the parents be satisfied with just one number reported for each of the two teams in order to have a complete picture? Record your responses as to Why or why not? What if the superintendent said he thought the 2 classes, which each had 25 students, were the about the same. Would you agree? The only scores reported were the team scores.(Teacher: Listen to the students' comments!)

Now to further confuse the interpretation of data, there are three types of averages. They are the mean, mode and median. Each is used by the business world, education, the news media, the financial world, weather, sports, and in the medical world to tell a story. The mean was defined above.

Definition: 30: The MODE for a set of data is the most popular or the most frequently occurring element or score in the set.

Definition: 31: The MEDIAN for a set of data is the middle element or score, when the elements or scores are arranged in order from lowest to highest.

What is the median and mode for each of the two math teams listed above?

Note: 80 (2019) and 91(2018) are the medians, **there is no mode.**

These three (mean, mode and median) are many times called the average, but the one that is being referred to is not always indicated. Each of these averages will be further clarified and compared in the applications.

Activity 7.2.2

Applications
(Some answers are given for your benefit and quick check.)

1. The following is a set of income figures representing the annual payroll in a small business, including the owner's salary.

    $18,000, $18,000, $18,000, $22,000, $25,000, $25,000, $80,000 Which do you think is the salary of the owner? Why?

    Answer: $80,000

2. Using Definitions 29,30, and 31 calculate the three averages for the data in #1. Use your calculator.

    The Mean is _____. The Median is _____. The Mode is _____.
    Answers: Mean = $29428.57. Median = $22000. Mode = $18000

3. The mean, mode, and median in #2 each were reported as the average!

    a. Which average do you think the owner claimed for the average pay? Why?
        Answer: Mean or $29428.57

b. Which average do you think the employees claimed for the average pay? Why?
Answer: Possibly the mode, $18,000.

4. What is the mode for each of the 2018 and 2019 math team scores?

2015 scores: 100, 99, 80, 76, 70
2014 scores: 100, 95, 91, 80, 59
Answer: There is no mode.

PLANE AND SOLID GEOMETRY ESSENTIALS

Home Activity 7.2.3
More Applications with answers

1. In the following calculate and identify the three averages. Use your calculator and/or computer. First arrange the scores in ascending order, and then calculate the three averages.

   a. 3,5,7   b. 0, 1, 3, 5, 16   c. 0, 1, 1, -2, -3   d, -3, 6, -7, -1, -4, 7
   e. 71, 65, 98, 57, 64, 69, 87, 88, 79, 48, 77, 80, 50, 75, 85, 90, 99, 30, 94, 96, 80.

   Answers:

   1a. Mean = 5, Mode = none, Median = 5
   1b. M = 5, Mode = none, Median = 3
   1c. M -3/5, Mode 1, Median = 1
   1d. M = 0, Mode = none, Median = -1
   1e. M = 1582/21 = 75.3, Mode = 80, Median = 79

2. Many times scores are grouped for easier calculating of the averages and for graphing. In the following case the scores are grouped to intervals of five and assumed the scores are at the center value.

   a. How many students took the test?
   b. What is the sum of all the scores? Assume the scores are at the center of the interval. 5(98) + 9(93) + 10(88) + ... + 3(53).
   c. What is the mode score?
   d. What is the median score?
   e. What is the mean score?
   f. Construct a graph to show the results
      96 -100 5 students   Scores: 96, 97, 98, 99,99.
      Counted as 5 Score at 98

10 students counted at a score of 93

15 students counted a score of 88

This method saves time but with the computer it is just as easy to put in the actual scores and be more accurate.

Answers: a. 99 b. 7552 c. 73 d. 78 e. 76.3

3. Graph (The students may bring in a variety of graphs.)

4. Calculate the Mean for the actual scores in the 96-100 and compare it with mean using the single figure or centralized score. 97.8 compared to 98

Challenge Activity 7.2.4

Write each of the first ten counting numbers using up to ten toothpicks or ten equal line segments for each number. This resembles the method used by digital clocks.

(1, 2, 3,4, 5, 6, 7, 8, 9, 10, 11, 12)

Example: To show the number 1 takes 2 toothpicks

Seven: takes 3 toothpicks and

Three: takes 5 toothpicks.

Write your summary for this session.

EVERYDAY DECISION MAKING
via
PLANE AND SOLID GEOMETRY ESSENTIALS

Chapter 7
Session 3
Decision-Making Methods Review
Jordan Curve Theorem

A mathematician, like everyone else, lives in the real world. But the objects with which he works do not. They live in that other place--the mathematical world. Something else lives here also. It is called TRUTH.

Jerry P. King
THE ART OF MATHEMATICS

The major objective is for you to understand that all decisions are based on undefined terms, defined terms, basic assumptions and previous decisions, laws or ordinances (theorems in geometry). The following will refresh and review the methods. These Methods for arriving at decisions range from guessing to formal logic. Everybody wants to make the correct and valid decisions! Conclusions using various methods for decision-making are:

Guessing
Illusions
Listening to experts or non-experts(both sides of an issue)
Observation
Induction
Estimating
Direct reasoning
Indirect reasoning

Statistics
Data collection Mathematics
Forms of an Implication
Ads

(The teacher may adapt the problems to fit local experiences to make the problems more meaningful using the students' backgrounds, activities, interests, and future plans. The way a problem is stated can motivate the students to investigate it.) The reason the model is Geometry has been used is that it is a very simple deductive system, one that all can understand at this stage of the "game", and also is practical in the trades.

Decision- Making Methods

Activity 7.3.1

Guessing:

This is a very weak way to make a decision, it amounts to basically tossing a coin to determine the decision. The only feature of it is you have passed the decision-making responsibility to the coin, but not the responsibilities of the outcome.

Illusions:

A method that uses what you think you see, which may not be the true case. The decisions may be based on what you see. This type is used by witnesses to an event.

Example 1: The following picture contains a photo of what a friend in Norway saw one summer and what another observers saw. One said a head of a duck and the other observer said the head of a elephant. What do you see?

From Elander's file

Example 2. The following example is one of my favorites. First look from the left and then from the right. (If you were a witness, what would you report having seen?)

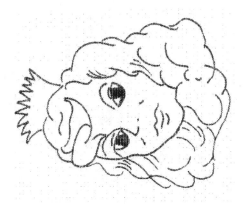

Puck Magazine 1915, W. Hill
From Elander's File

This was also sent from a Norwegian Friend who labeled it young Lena or Grandma Lena.
I believe it is really from Puck magazine 1915 by W. Hill

Example 3. Have you seen the arrow on the FedEx trucks? Once you see it you will never forget it and every time you see the truck you will see the arrow.

FedEx

Observation and Induction Reasoning

Inductive Reasoning: This type arrives at a conclusion after investigating a few cases. This type requires record keeping to detect a pattern or trend.

Teacher: Be sure to do these two examples!

Example 1: How many straight chords or segments can be determined by n points on a circle?

a. Draw a circle with a 2 inch diameter.
b. Locate 2 points on the circle and draw the chord determined by the two points. Conclusion: 2 points, one chord.
c. Place another point on the circle and now predict the number of chords. Then actually count the number and try to complete the table for the relationship between the points and number of chords.

Points	Chords
2	1
3	3
4	6
5	?
6	?
7	? conclusion Chords = $n(n-1)/2$

This solution (by induction) seems to give the answer.

Example 2: Repeat example 1, but instead of chords count the regions and try to see if there is a pattern for prediction. You will be surprised!

Points	Regions
2	2
3	4
4	?
6	?
7	?

Inductive Reasoning is drawing a conclusion as to future events from a few past events. This method for arriving at a conclusion is like predicting the future and is questionable. The medical profession, weather forecasting, financial investors, auto repair, and other trades and professions use this type of reasoning to reach a conclusion. Ask students to explain the how inductive reasoning is used for each indicated case above.

A different type of example of an induction problem.

Example: A farmer feeds a turkey every morning for months, but on Thanksgiving Day or the day prior the farmer comes out to the feeding area and the turkey is in for a disappointment. What happened? The turkey made a false conclusion using inductive reasoning.

Many times conclusions may not be valid following from the given statements.

1. The junior class in a school A voted on the following question: Should cell phones be turned off in classes? (Only 15% voted no.) Which of the following statements printed in 5 different papers is the most correct? Good discussion question!

    a. High School students vote to turn off cell phones in class.
    b. 85% of high school students vote to use cell phones in school.
    c. Most students don't use cell phones in classes.

d. 15% of students use cell phones in classes.
e. The junior class in school A votes to restrict the use of cell phones in classes.

Remember: Statements are true or false and conclusions are valid or invalid.

2. **Fact**: All students of school X, wear red caps at the football games.

   Fact: John is at the game and is wearing a red cap.
   Conclusions: Which are valid?

   a. John must be a student of school X.
   b. Peter is not wearing a red cap at the game, then he is not a student at X.
   c. Joe is not a student at X, then he does not wear a red cap at the game.

   Listening to experts or non-experts. (Tv, Radio, media may not state the source of their opinions.)

3. The problem with this method is that issues are restated in various ways and often misinterpreted as to the truth of the statements and conclusions. The advertising world and other "worlds" (such as politics) use these forms to possibly take advantage of people who do not understand them and the issues.

   The forms of implications are: A   B

   Original statement: If A, then B. (valid)
   Converse: If B, then A. ?
   Inverse: If not A, then not B. ?
   Contrapositive: If not B, then not A.(valid)

4. **Fact**: If you fly the Flag on Memorial Day, then you are patriotic. Which of the following are valid?

   a. If you are patriotic, then you fly the flag on Memorial Day.
   b. If you do not fly the flag on Memorial Day, then you are not patriotic.
   c. If you are not patriotic, then you do not fly the flag on Memorial Day.

5. **Fact**: If you watch TV-PBS, then you want factual information. Which of the following may be valid? Explain your answers.

   a. If you don't watch TV-PBS, then you don't want factual information.
   b. If you don't want factual information, then you don't watch PBS.
   c. If you want factual information, then you watch TV-PBS.

Data collection

6. Polls are many times used to indicate a trend. Listed are questions you should know about polls. Give an example of each of the following and how it can influence the result. (Information related to the following is very seldom reported.)

   a. The number polled. (very seldom reported.)
   b. How collected, time of day, the questions. (randomness, age, location)
   c. How contacted, email, telephone, U.S. mail, or interview.

(It has been reported that data is now collected from the social media. Do you think that method is random or from experts?)

Teacher: 1. Suggest students bring information or examples as to the above from various sources!
        2. Research and report the famous poll taken to predict the winner of the 1936 U.S. Presidential election. (This is very interesting and a must! Parents would be interested in it also.)

The last Word

Decision-Making was the reason Socrates and Plato started their academies for futures leaders, which led to the decision for requiring geometry for all students, but the assumption was and still is that a student will become a valid decision maker by taking geometry. But it was the NCTM 13th yearbook (Nature of Proof) that justified that the geometry course must be taught using everyday decision-making situations in order to produce valid decision makers.

Teacher: See Fawcett's NCTM 13th yr. Bk, NATURE OF PROOF.

The other reason for teaching geometry to all students is its practical applications in so many areas. (Let the class list some areas)

An Interesting theorem

Theorem. 44: Jordan Curve Theorem states that if an area is separated by a line, then if you cross the line an odd number of times you are on opposite sides, and if you cross the lines an even number of times, then you are on the same side as you were at the beginning.

Application: Carnivals or State Fairs many times have a house or a maze which you are to enter and find your way out. To easily solve (find you way out) keep your right or left hand on the wall and it will guild you to the exit. This method may not be the shortest, but it will lead you to the exit. Try this method on the following maze.

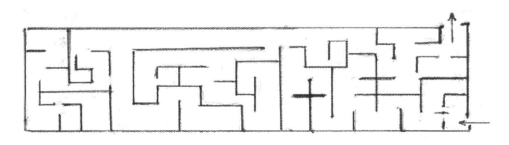

EVERYDAY DECISION MAKING
Via
PLANE AND SOLID GEOMETRY ESSENTIALS

Chapter 8
Session 1
Review of Decisions-Making Methods

A mathematician, like everyone else, lives in the real world. But the objects with which he works do not. They live in that other place--the mathematical world. Something else lives here also. It is called TRUTH.

Jerry P. King
THE ART OF MATHEMATICS
From Elander's File

Rodin's THE THINKER
California Golden Gate State Park

More review of the major points of
EVERYDAY DECISION MAKING.

The major objective is for you to understand that all decisions are based on undefined terms, defined terms, basic assumptions and previous decisions, laws or ordinances (theorems in geometry). The following will refresh and review the methods people use in Decision Making.

The methods for arriving at decisions range from guessing to formal logic. Everybody wants to make the correct and valid decisions! Conclusions from various methods of decision-making used by people are:

Guessing

Illusions

Listening to experts or non-experts(both sides of an issue)

Observation

Induction

Estimating

Direct reasoning

Indirect reasoning

Statistics

Polls

Forms of an Implication

Ads

If the teacher can adapt the above cases to fit local experiences to make the conclusions more meaningful and valid for the students' background, activities, interests, and future plans. (The way problems are stated can motivate the students to investigate it.) The reason the model of Geometry has been used is that it is a very simple deductive system at this stage. One that all can understand. This basic method for decision making was first created and used by Socrates, Plato, Pythagoras, Euclid, Archimedes, Descartes plus many others, and even now it is still required in many programs at the universities. It has been required in the high schools the last 150 years. The course in Geometry has two objectives: one is the use of geometry in the various trades and professions, the second is the teaching for better Decision

Making. In most cases the second reason has been assumed as a result, but Prof. Fawcett verified that to achieved this objective of Decision Making, it must be taught with applications from everyday situations.

Review the Methods

Guessing:

This is a very weak way to make a decision, it amounts to basically tossing a coin to determine the decision. The only feature of it is you have passed the decision-making responsibility to the coin, but not the responsibilities of the outcome.

Illusions:

A method that uses what you think you see, which may not be the true case. This type is used by witnesses to an event.

Example 1: The following picture is a photo a person took and what two observers saw. One said a frog and the other a horse. What do you see? (Look from the left and then from the right.)

A friend sent the above picture to the author.
From the Elander's File

Example 2:   The following picture of the Canadian Flag has the profiles of two faces with pointed noses pointing at the base or stem of the leaf. Hint: The **large nose and open mouth** is profiled. Do you see the two outlines of the two faces?
(Focus on the white!)

(What are the colors of the Canadian flag?)

From Elander's file

Observation and induction

Inductive Reasoning: This type arrives at a conclusion after investigating a few cases. This type requires record keeping to aid in detecting a pattern or trend.

A previous inductive reasoning problem relating points on a circle to the number of chords. You may want to review it.

Example 2: A coin was tossed 7 times and a head came up 7 times. What would be your prediction for the next toss? (Class discussion!)

Example 3: (An old sailor's saying.) Wind from the East is not good for man nor beast. How do you think they arrived at the saying?

Direct reasoning: Arriving at a conclusion from given valid information

Example 1: Your State Law states a person is to have a valid driver's license to drive. John drives a car. What is your conclusion?

Teacher: A → B is the law. The students will probably state the converse.

Example 2: Your State requires a person to have a valid driver's license to drive. John has a driver's license. What is your conclusion?
Teacher: listen to the conclusions.

Example 3: All students of Z high school wear red caps at the school's sporting events. John is wearing a red cap at the game, therefore he attends Z school. Valid or not valid.

Indirect reasoning, Recall examples:

Example 1: Assume: The square root of 2 can be expressed as an integer over an integer. This assumption is false by Indirect Reasoning.

Cases: $\sqrt{2}$ = o/e, or e/o, or o/o and show that each of these cases is impossible.

Therefore .....

Example 2. Galileo's conclusion was that objects heavier than air fall at the same rate. Assume that heavier object fall faster, (This was the assumption at that time.)
His reasoning: Tie the 2 objects together and they should fall faster, but shouldn't the lighter object slow down the heavier object? Hence the contradiction leads to the conclusion that they fall at the same rate.

Statistics:

Example 1. A teacher was asked by the principal to provide some information as to the membership in his club to justify the club for next year. The teacher is the sponsor of the club. Here is the information the teacher provided. The club initially had x members, one member quit, but we now have 2(x-1) or twice as many members. Should the school principal approve the club or not?

The facts: Initially the club had only 2 members. Does this now change you answer? Explain your reasoning?

Polls: Conclusions obtained by taking a poll should:

a. State where and how the information was collected.
b. Time of day the data was collected.
c. The number that was polled.

Question 1: Should students in your school be permitted to use c- phones in classes. Would the responses be different if you polled only the students who have phones? Or only the teachers? Only the parents?

Example 1. In the 1936 presidential election a nation wide poll was taken and it predicted a landslide victory for Alf Landon. The election was actually a victory for FDR. Why was the poll wrong? The poll was taken via the telephone and hence the poll sample was not random. Who had telephones? (Your parents may enjoy this!)

Implications

Student's should know the forms of an implication. Theorem, Converse, Inverse, and Contrapositives.

Theorem and Contrapositive are valid and converse and inverse may be valid. Business Ads usually only provide one sided information.

Example: In the 1940 and fifties a smoking ad was: Try our product for 30 days and then if you don't like it quit!
Sounds fair enough, but in 30 days you may be so addicted you can't quit.

Optional

The game of Sprouts (A class activity)
Suggestion: Teach the students how and suggest they teach their parents.

This game uses your skill to look at former results and predict the out come (inductive reasoning). It is a one, two or three-person game and is called sprouts, Thanks to Martin Gardner, who gives credit to other mathematicians. The simplest game is played as described below. The objective is to be able to **predict** the winner, the number of moves by each player, and the total number of moves in the game given the number of original seeds.

Rule 1: A seed can only have 3 stems coming from it.
Rule 2: At the midpoint of a stem you add a new seed.
Rule 3: A seed is dead if it has three stems on it.
Rule 4: The player to make the last possible move is the winner

The simplest game consists one seed and 2 players (This case is played out for you so you will understand the moves and the winner,

1 seed
0

Players A and B seed   seed
O

A has the first move.      Result:

Second move is by B. B's turn and B can add a stem from the left seed to the right seed and put a new seed at the midpoint.

Now it is A's move and there is no possible move by Rule 3. So B wins.

Conclusion: **Staring with one seed and 2 players the second player wins. There was only 2 moves in the above game.**

The next game now begins with 2 seeds and 2 players. It is advisable for each player to use a different colored maker which will make the counting for the number of turns much easier. The winner is the person makes the last move.

When a seed has three stems it is dead (rule 1) and no more stems can be connected. Here are some illustrated moves. And results

Player	seeds	Moves	winner	total moves
1	2			

Move 1   1   1

                           1   2

Player 2, Move 2   do you see the dead seed?

(After 4 moves, there will be a winner)

Now it is 1 turn

In move 3 there are three dead seeds, but there are more possible moves. Continue until no more possible moves. Conclusion: Who wins and in how many moves? This game above will end after 5 moves and the winner will be the person who made the first move

Summary for game starting with 2-points

Results: Players 2 Seeds 2 Moves 5 Winner Player 1
(Like most math games, it gets harder as the game advances.)

In this game of sprouts the objective is to predict the winner and the number of moves, given the number of starting seeds and 2 players.

Teacher: Suggest the students play the game at home and teach their parents and friends.

Again: Inductive Reasoning is drawing a conclusion as to future events from a few past events.

This method for arriving at a conclusion is like predicting the future and is questionable. The medical profession, weather forecasting, financial investors, auto repair, and other trades and professions use this type of reasoning to reach a conclusion and many every day predictions.

A different type of example of an induction problem.

Examples: Calculator problem:

1/9= ? 2/9 = ? 3/9 =? 4/9 = ? 5/9=? 6/9 = ? 7/9 = ? 8/9 =?

Now predict the answer to 9/9 =? Then check the answer with the calculator. Many times, the conclusions may not follow from the given statements or cases.

Case 1. The junior class in a school A voted on the following question: Should cell phones be turned off in classes? (Only 15% voted no.)

Which of the following statements printed in 5 different papers is the most correct in your opinion and why?

a. High School students vote to turn off cell phones in class.
b. 85% of high school students vote to use cell phones in school.
c. Most students don't use cell phones in classes.
d. 15% of students use cell phones in classes.
e. The junior class in school A votes to use of cell phones in classes.

Remember: Statements are true or false but conclusions are valid or invalid.

The forms of implications are:

Original statement: If A, then B. (valid and T?)
Converse: If B, then A. (Valid? and T?)
Inverse: If not A, then not B. (Valid? and T?)
Contrapositive: If not B, then not A. (valid and T?)

4. Fact: If you fly the Flag on Memorial Day, then you are patriotic. Which of the following are valid?

   a. If you are patriotic, then you fly the flag on Memorial Day.
   b. If you do not fly the flag on Memorial Day, then you are not patriotic.
   c. If you are not patriotic, then you do not fly the flag on Memorial Day.

Data collection

6. Polls are many times used to indicate a trend. Listed are questions you should know about polls. Give an example of each of the following and how it can influence the result. (Information related to the following is very seldom reported.)

   a. The number polled. (very seldom reported.)
   b. How collected, time of day, the questions. (randomness, age, location)
   c. How contacted, email, telephone, U.S. mail, or interview. (It has been reported that data is now collected from the social media. Do you think that method is random or from experts?)

Teacher: Suggest students bring information or examples as to the above from various cases!

The last Word

Decision-Making was the reason Plato started his academy for future leaders, which led to the decision for requiring geometry for all students, but the assumption was and still is that a student will become a valid decision-maker by taking geometry. **The geometry course must be taught using everyday situations in order to produce valid decision makers!**

(Teacher: See Fawcett's NCTM 13th yr. bk: NATURE OF PROOF)

Write your summary and comments.

GEOMETRY
PLANE AND SOLID
ESSENTIALS

Chapter 9
Descartes' Gift

The important thing is to not stop questioning...

Albert Einstein
From Elander's file

You have no doubt noticed that most of the geometry you have studied is really old, but, in the 1600s, Rene Descartes created what could be called the beginning of modern mathematics. I am sure you would like a peek at his creation! It could be stated as a union of geometry and algebra. He started with the idea that a point on a sheet of paper can be defined as a set of ordered pair of numbers. You have already done this when graphing points in algebra where you were taught how to plot points. Example locate the point (2,3) on graph paper.

$$
\begin{array}{ccc|cccc}
 & & & y & & & \\
 & & 2 & & & & \\
 & & 1 & & & & \\
-3 & -2 & -1 & 0 & 1 & 2 & 3 \\
\hline
 & & -1 & & & & \\
 & & -2 & & & & \\
\end{array}
$$

As an activity locate the following curves (use graph paper) and name the figures. (Use your calculator)

1. Connect the points (-3,0), (0,3) and (3,0) and name the figure.

2. Graph the equation $y = x$ and $y = -x$ and name the result.

   (For values of x greater than -6 and less than +6 for integers and connect the points)

3. Graph $y = (1/2) x^2$ for x values for -6 to 6 and connect the points.

   Can you name the curve?
   Draw the curves

4. Graph $x^2 + y^2 = 100$ for values of x from -10 to +10. Can you name the curve?

5. $x^2/16 + y^2/9 = 1$ for values of x from -4 to 4, Name the curve.

6. $Y = Sin X$ for X angle values from 0 to 180 every 10 degrees.

   (Teacher should aid the students)

ESSENTIALS
FOR
PLANE AND SOLID GEOMETY

Chapter 10
Epilogue

Congratulations, you have completed a program that will improve your test scores and everyday Decision-Making skills via Plane and Solid Geometry Essentials. (Geometry and basic decision-making has been reported to be the poorest results on college entrance exams.) This is why basic Decision-Making Skills was incorporated into the Geometry Essentials program. These sessions helped you understand the key theorems in 2-D and 3-D Geometry, plus to provide some worthwhile experiences, not only in the areas of applied mathematics, but in the world of everyday Decision-Making. All thinkers understand that conclusions are based on **undefined terms, defined terms, assumptions, and theorems or laws previously proved**. You should now understand why Plato had posted at his adult school's entrance:

LET NO ONE IGNORANT OF GEOMETRY ENTER HERE.

Plato (600 BCE)

Note: Plato and others as you now understand, did not mean just the study of geometric points, lines and planes, plus geometric figures, but he meant the training in logical thinking resulting from the study of geometry. That all conclusions involve undefined terms, defined terms, basic assumptions and former conclusions. He and Socrates knew that a democracy (Athens was the first) to function efficiency depended on an informed critical thinking voting public. The author has included bits of history to make the details more interesting. Euclid (300 BC) is credited with writing the first geometry book in the logical order to which Plato wanted his future leaders to learn from. The following is an interesting cover for Euclid's geometry text translated by Henry Billingsley in the

1570s. What is your interpretation from the cover? Look at all the figures and titles. This is another reason for teachers to participate in professional organizations. Study the cover, it is very revealing as to the various topics at that time!(Show it to your parents)

It provides reasons why geometry is studied.

The author's final statement is:

LET NO STUDENT IGNORANT OF GEOMETRY EXIT HERE.

Keep in mind that you have only been exposed to the "tip" of the mathematical iceberg. Like the visible part of an iceberg, only a small portion of is above water. There is a vast amount of math still uncovered for you. Perhaps you will take another course in the future, or one in Math History.

I sincerely hope you can say that this "Geometric Adventure" was worthwhile, useful and even enjoyable at times.

Respectfully,
Jim Elander (Author)
Retired in beautiful Montana
Naples, FL

Email: jelander@aol.com or elanderje@gmail.com

Teacher: It is recommended that the post-test be given in a class period and scored the next class period. The scores for each student should be compared with their score form the pre-course test they took at the beginning of the course and can now be returned.

Copy the pre-course test and use it! (page xv)

Answers to Post Exam: Number of answers is 42? Compare your score with the test score on the Entrance Exam. You should be satisfied!

1. $50 < x < 200$
2. 1 5 10 10 5 1
3. 360 degrees
4. 4 planes
5. 6 segments, 10 segments, $n(n-1)/2$
   undefined is between 6 and 9. The two that probably need defining are "good" and "aid."

6. 7
7. No

8. 5 units
9. True
10. Yes
11. Adjust the gate to form a rectangle and then add a diagonal.
12. One (1)
13. Check for perpendicularity from two directions
14. proven important mathematical statement.
15. An Assumption
16. P is 40 ft. Area of box is 100 sq. ft.
    Circumference (circle) 10(3.14) = 31.4 units
    (circle) is 25 (3.14) = 78.5 sq. units
17. a. Vol. of box is 240 cu. ft ?
    b. Vol. of cylinder is 785.4 cu. cm.
    c. Vol. of ball is 33.5 cu. in.
    d. Box area is 248 sq ft. Cylinder area is 314.2 sq cm.. Baseball area is
        50.3 sq. in.

18. "c" is valid.
19. What is the mean, median and mode for the following?.

	AM		AM	PM
A	*****		A	***
B	****		B	******
C	*		C	**

20. a. Draw graph for AM and PM classes in #19.

    Number of answers is 39. (The teacher may wish to count the number of
    answers differently.)

Your pre-score_____ post-score(C)_____

Are you satisfied? Round C (post score) to nearest whole number.

**Teacher: Also return the pre-course scores to the students to show their
        progress to their parents.**

PLANE AND SOLID GEOMETRY ESSENTIALS

Appendix 1
Definitions

(It is assumed some of the geometric terms, which are classified as undefined have been introduced in previous years are not necessary since they have been covered in previous math classes.)

Such as:

Point, ray, line, and plane are classified as undefined terms

Point ●

Ray ● ⟶

If you move a point continually in one direction, a ray is generated. (Sometimes called a half line.)

Line ⟵————————⟶

If you move or roll a point continually, both directions, a line is generated. In geometry, a line means a straight line.

Plane

A plane in boundless, so the above is called a plane segment.
A plane is generated by moving a line segment.

Line segment ●————————●

If you roll or move a point a set distance, then a line segment is generated.

Tools

A ruler or straightedge, protractor, and calculator are the basic tools, even a computer could be added now days.

Definition 1: A logical system is one based on Undefined Terms, pg. 3
Defined Terms, Assumptions, and previously proven or accepted Conclusions. Example: If A=B and B=C then A=C

Definition 2: A theorem is an important mathematical statement pg. 3
which has been proven.

Definition 3: A definition is valid if it is true when reversed. pg. 3
(converse)

Definition 4: Parallel lines are lines on a plane that do not pg. 18
intersect no matter how far extended.

Definition 5a: A triangle is the set of three non-collinear points pg. 18
and the line segments determined by the three points.

Definition 5b: Angle consists of two rays with the same beginning pg. 18
point.

The method for labeling angles is by using three letters, one letter on each ray and one at the beginning point. The beginning point is called the vertex.

Example: ∠ADG where the middle letter is always at the vertex. Notice the symbol for angle (<).

The drawing for ∠ADG is:

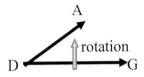

The following were reviewed in Session 4:

An angle of 360 degrees (symbol for degree is °) in a complete rotation or circle.

A straight angle or half circle is 180°.
A right angle is 90°.
1° equals 60 minutes (symbol for minutes is ´.)
The first set of minute parts.
1 minute equals 60 seconds (symbol for seconds is ˝.)
The second set of minute parts.

Definition 6: Direct proof is a conclusion that follows logically.

Definition 7: Indirect proof is a method where you list all the possibilities and show that all but one is impossible, therefore the remaining possibility is the correct one.

Definition 8: Two polygons figures are similar, if when mapped the corresponding angles are equal and the ratios of the corresponding parallel sides are equal.
The symbol that indicates the figures are similar is ~, like a lazy s.

Definition 9: A polygon is a closed figure consisting of n points and the line segments determined by the n points, such that none of the line segments intersect except at their end points.

Definition 10: A triangle is a three-sided polygon.

Definition 11: Congruency: If two polygons are similar and the ratio of the corresponding sides is1, then the fi

        Note: In Session 4: Q.E.D (quod erat demonstrandum) is at the end of many proofs and means: "That which was to be proved, has been."

Definition 12: A parallelogram is 4 sided polygon, where the opposite sides are parallel.

Definition 13: A rectangle is a parallelogram with right angles.

Definition 14: The Area of a geometric figure is the number of square units contained in the interior of the figure.

Definition 15: A trapezoid is a quadrilateral with only two sides parallel. Session 5: Also explains:

    a. Triangles types are also acute, right, or obtuse.

    b. A triangle is a rigid figure and its importance is in construction.

Teacher: The terms trapezoid, perimeter, altitude, and the English and Metric systems of measurement are used.

Definition 16: In a right triangle, the side opposite the right angle is the called the hypotenuse.

Definition 17: In a triangle, the altitude is the line segment from the vertex of the angle perpendicular to the opposite side. (The side may have to be extended.)

Definition 18: In a triangle, a median is defined as the line segment from the vertex of the angle to the midpoint of the opposite side.

The four forms of an implication "If A, then B." are:

| Theorem | If A, then B. |
| Converse | If B, then A. |

| Inverse | If not A, then not B. |
| Contrapositive | If not B, then not A. |

In Session 7: The students reviewed the following terms, triangle, polygon, parallelogram, rectangle, trapezoid, area, altitude, perimeter, square, also altitude, median and angle bisector related to a triangle. The four forms of an implication are use in the exercises.

Definition 19: A circle is defined as a plane figure consisting of the set of all points on the plane that are a given distance from a given point. (The given point is labeled the center and the given distance is called the radius.)

Definition 20: Circle terminology
   a. The given point in the above definition is called the center.
   b. The line segment from the given point (center) to a point on the circle is called the radius.
   c. The distance around the circle is the circumference.
   d. The line segment from one point on a circle to another point on the circle is called a chord.
   e. If a chord contains the center point, it is called a diameter.
   f. A line that intersects a circle in only one point is called a tangent.
   g. A secant is a line that intersects a circle in two points.
   h. A portion (segment) of a circle is called an arc.

Definition 21: A circle's central angle is an angle with its vertex at the center of the circle.

Definition 22: The measure of the arc a central angle intercepts is the same measure as the central angle.

Definition 23: An inscribed angle is an angle where the vertex is on the circumference of a circle and the rays or sides are chords.

Terms introduced:

Triangles types and terms reviewed: acute, obtuse, and right, median, Center of Gravity, altitude. Concept: 3-D View and Layout

Views	3-D View	Layout view

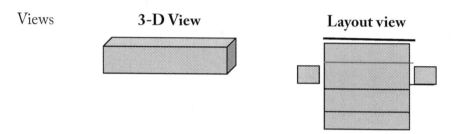

Definition 24: Volume is the number of cubic units a 3-D figure contains.

In Sessions 14-16: The formal definitions are NOT stated but meaningful ones are introduced via explanations and in applications for, prism, cylinder, cone, and pyramid and the formula for volume and surface areas are justified.

Definition 25: A sphere is the set of all points equidistant from a given point, called the center.

Definition 26: The SINE (Sin) of an acute angle in a right triangle is the ratio of the length of the side opposite the angle divided by the length of the hypotenuse.

Definition 27: The COSINE (Cos) of an acute angle in a right triangle is the ratio of the length of the adjacent side divided by the length of the hypotenuse.

Definition 28: The Tangent of an acute angle in a right triangle is the ratio of the length of the side opposite the angle divided by the length of the side adjacent to the angle.

These Terms were introduced.

Latitude, circle of Latitude
Trig form of the Pythagorean Theorem
Sin and Cos curves
Slant height
Diagonal of a cube

Terms and forms of an implication are used:

Name   Symbolic
Statement: A --> B
Read: If A, then B.
Converse: B --> A
Read: If B, then A.
Inverse: ~A-->~B
Read: If not A, then not B. The negation symbol, read not is~.
Contrapositive: ~B -->~A
Read: If not B, then not A.

Validity and Truth of Statements are illustrated.

Definition 29: The MEAN or arithmetic average for a set of
               numbers is the sum of the numbers divided by n,
               the number of numbers in the set.
               Formula: Mean = (sum of n scores)/n.
Definition 30: The MODE for a set of data is the most popular
               or most frequently occurring element or score in
               the set.
Definition 31: The MEDIAN for a set of data is the middle
               element or score when the elements or scores are
               arranged in order from lowest to highest.

GEOMETRY ESSENTIALS

Appendix 2
Postulates

Post. 1: A line has an infinite set of points.

Post. 2: Two points will determine one and only one straight line.

Post. 3: There is a one to one correspondence between the points on a line and the real number line.

(Postulate 3 enables you to use algebra to help solve geometric problems.)

Post. 4: The shortest distance between 2 points is the measure of the straight line segment determined by the two points. (This is not always true as you saw in the taxicab exercise in Session 1, and that is why it is a postulate.)

Post. 5: The shortest distance between a point and a line (on a plane) is the perpendicular line segment.

Post. 6: Three non-collinear points can determine one Geometric plane.

Post. 7: If two parallel lines are crossed by another line (called a transversal), then the alternate interior angles are equal.

Post. 8: Through a point not on a given line there is only one line through the point that is parallel to the given line.

Post 9. If two triangles have two corresponding angles equal, then the triangles are similar. (AA)

Post. 10: If 2 triangles have their corresponding sides in equal ratios, then the triangles are similar. (SSS)

Post 11. If two triangles have two corresponding sides in equal ratio and the included angles equal, then the triangles are similar. (SAS)

Post. 12: The area of a rectangle is equal to the length times the width and The answer is in square units. (The basic unit of measure for base and height must be the same.) Formula: $A = bh$ sq. units.

Post. 13: The formula for the circumference of a circle is $C = 2\pi r$ or πd.

Post. 14: The formula for the area of a circle is $A = \pi r^2$.
(Postulates 13 and 14 are proved in Chapter 6.)

Post. 15: The volume of a 3D figure is the number of cubic units it contains by multiplying K (a constant) times the area of the base times the height.

Postulates for the real numbers system

1. Identity postulates for operations with numbers.
   $a + 0 = a$ and $a \times 1 = a$

2. Closure postulates for operations of addition and multiplication
   $a + n = N$ and $a \times n = N$ $a/n = N$ (n cannot be 0)

3. Commutative postulates for operation of addition and multiplication.
   $a + b = b + a$ $a \times b = b \times a$

4. Associative postulate:
   $a + (b + c) = (a + b) + c$
   $a(bc) = (ab)c$

5. Distributive Postulate: $a(b+c) = ab + ac$

PLANE AND SOLID GEOMETRY ESSENTIALS

Appendix 3
Theorems

Th. 1: In a triangle the sum of two sides is greater than the third side, and the third side is greater than the difference of the two sides.

Th. 2: If two lines intersect, then the opposite angles are equal.

Th. 3: The sum of the angles in a plane triangle is 180 degrees.

Th. 4: The exterior angle of a triangle is equal to the sum of the two non- adjacent interior angles.

Th. 5: If two lines a and b are crossed by another line (transversal) so that the alternate interior angles are equal, then lines a and b are parallel.

Th. 6: In an isosceles triangle, the angles opposite the equal sides are equal in measure.

Th.7: In an equilateral triangle, all the angles are equal.

Th. 8: In a scalene triangle, there is a correspondence between the length of the sides and the angles opposite the sides. The larger the side, the larger the opposite angle.

Th. 9a: If the polygon is a Parallelogram, then the diagonals bisect each other.

Th. 9b: If the figure is a parallelogram, then the opposite angles are equal and the opposites sides are equal in.

Th. 9c: If the figure is a rectangle, then the diagonals are equal.

Th. 9d: If the figure is a parallelogram and the diagonals a e perpendicular, then the figure is a square.

Th. 9e: If the diagonals of a 4 sided polygon bisect each other, then the figure is a parallelogram.

Th. 9f: Converse of 9a.

Th. 10: The area of a triangle is ½ the base times the height or ½ base times the altitude. Formula: $A\triangle$ = (½)bh sq. units.

Th. 11: The area of a parallelogram is equal to base time the height, or base times the altitude. Formula: A = bh sq. units

Th. 12: The area of a trapezoid is the sum of the two bases times ½ the the measure of the altitude.

$$\text{Formula is } A = (½)h(b_1 + b_2) \text{ sq. units}$$

Th. 13: Any point on the perpendicular bisector of a line segment is equal distant from the end points of the segment.

Th. 14: If a point is equal distance for the end points of a line segment, then it is on the perpendicular bisector of the segment.

Th. 15: If a point is on the bisector of an angle, then the point is equal distance from the rays of the angle.

Th. 16: If a point is equidistance for the rays of an angle, then the point is on the angle bisector. Converse of Th. 15.

Th. 17: In a right triangle with sides a, b, and c is the hypotenuse, then a2 + b2 = c2. VIT (very important theorem) (Th. 41. Trig form of the Pythagorean Theorem. $(Sin A)^2 + (Cos A)^2 = 1$)

Th. 18: In a 30-60 degree right triangle the side opposite the 30 degree angle is angle ½ the hypotenuse.

Th. 19: In 30-60 degree right triangle the side opposite the 60 degree is half the hypotenuse times the $\sqrt{3}$ or $(h/2)\sqrt{3}$.

Th. 20: The three altitudes of a triangle intersect at a point (are concurrent), but the sides may have to be extended.

Th. 21: The angle bisectors of a triangle are concurrent at a point equal distant from the sides of the triangle.

Th. 22: The three medians of a triangle are concurrent at a point 2/3 the length of each median.

Th. 23: The segment from the center of a circle to the point of tangency is perpendicular to the tangent.

Th. 24: The tangent from a point outside the circle squared, plus the radius squared, equals the square of the distance from the point to the center of the circle.

Th. 25: The perpendicular bisectors of the chords in a circle intersect at the center of the circle.

Th. 26: An inscribed angle in a circle is equal to ½ the degree measure of its central angle or ½ the degree measure of the intercepted arc. (See Def. 12.)

Th. 27: The volume of prism or cylinder is the product of the base area times the height or altitude, $V = A(h)$ cu. units. (Height or altitude is the measurement of the perpendicular distance. The measurements must be in the same units.)

Th. 28: The volume of a pyramid is the area of the base times the height or the formula is $V_{Pyramid} = Bh/3$ cu. units (B and h in the same units)

Th. 29: The formula for the lateral area of a pyramid is the sum of the areas of the triangles or A_L = (1/2)altitude of the triangles times the perimeter of the base. $A_{L=}$ (½) P(h)

Th. 30: The total area of a pyramid is the sum of the Lateral area plus the base area.

Th. 31: The volume of a cone is 1/3 the area of the base times the height or altitude. The formula is $V_{Cone} = \pi r^2 (h)/3$. cu. units).

Th. 32: The formula for the lateral area of a cone is A_L = (1/2) CSh sq. units, where C is the Circumference of the base and Sh is the slant height.

Th. 33: The formula for the total area of a cone is the lateral area plus the area of the base or TA_{Cone} = (1/2)CSh sq + $\pi\, r^2$ sq units.)

Th 34: The surface area of a sphere is $4\pi r^2$ sq. units, where r is the radius.

Th. 35: The formula for the volume of a sphere is $(4/3)\pi r^3$ cu. units, where r is the radius.

Th. 36: If a line segment (Like a flag pole) is perpendicular to two lines on the plane at the point of intersection, then it (the pole) is perpendicular to the plane.

Th. 37: (Archimedes favorite theorem) If a sphere and a cone are inscribed in a cylinder such that the radius of the cylinder, sphere, and the cone is R and the altitude of the cylinder and cone is 2R then the ratios of their volumes are 1:2:3, meaning the volume of the sphere is twice the volume of the cone and the volume of the cylinder is three times the volume of the cone. Ratio of the volumes is 1:2:3!

Th 38: Sin theorem is Sin A/a =SinB/b= Sin C/c in a triangle.

Th. 39a: Cos Theorem is $c^2 = a^2 + b^2 - 2abCosC$.

Th. 38b: Another form of the Cos Theorem is
$CosC = (c^2 - a^2 - b^2)/(- 2ab)$.

Th. 40: One formula for Pi (π) is the limit of n(sin 180/n) as n gets very large. Where n is the number of sides in the inscribed regular polygon.

Th. 41: $C = 2\pi r$

Th. 42 $A = \pi r^2$

Th. 43: Trig form of the Pythagorean Theorem, $(Sin\ A)^2 + (Cos\ A)^2 = 1$

Implications are very important and often times misinterpreted.

The four forms of an implication are are easily interpreted by Venn diagrams.

Statement	Conclusion
Theorem (A-->B)	True, and valid
Converse (B-->A)	Not necessarily true or valid
Inverse (~A-->~B)	Not necessarily true or valid
Contrapositive: (~B-->~A)	True and valid.

Remember: **Statements are classified as true or false, but conclusions are valid or invalid!**

An interesting theorem

Theorem 44: The Jordan Curve Theorem says if an area is separated by a line, then if you cross the lines an odd number of times you are on opposite sides, but if you cross the lines an even number of times then you are on the same side.

Remember the hand rule!

Appendix 4
Conversion Information

English		Metric or International (SI)
1 inch	=	2.54 centimeters (cm.)
1 foot = 12 inches	=	30.48 cm.
3 feet = 1 yard	=	.9144 m. or 91.44 cm

39.37 inches = 1 meter = 10 decimeters (dm.) = 100 cm.

5280 ft. = 1760 yds = 1 mile = 1.609 kilometers = 1609 meters

Radius of earth = 3963 miles = 6377 km. **1 nautical mile = 6075 yards**

Area (Approximations)

1 square inch	=	6.45 sq. cm.
1 square foot = 144 sq. in.	=	929.03 sq. cm.
1 square yard = 9 sq. ft.	=	5625 sq. cm. or .863 sq. meters
10.76 sq. ft. = 1.196 sq. yds	=	1 sq. meter = 10,000 sq. cm.
2.47 acres = 1 hectare	=	1 hectare = 10,000 sq. meters

1 acre = or 4,840 sq. yds. = 4,177 sq. meters = .4177 hectares

Volume approximations

1 cubic inch	=	16.39 cu. cm.
1 cu. ft. = 1728 cu. in.	=	.028 cu. meters
1 cu. yd = 27 cu. ft.	=	.765 cu. meters
35.31 cu. ft = 1.31 cu. yds.	=	1 cu. meter
1 quart	=	.9463 liters
1 gallon = 4 qts.	=	3.785 liters
1.0568 U.S. gallons	=	4 liters or 1 Canadian gallon

1 cubic foot of water is 62.43 lbs. or 28.31 kilograms

1 gal of water weighs 8.35 lbs.

Angle Measurement

1 minute = 60 seconds

1 degree = 60 minutes	=	.01745 radians
57.2957 degrees(rounded)	=	1 radian
180 degrees	=	3.1416 radians (rounded)

A few not so common measurement conversions

1 barrel = 31.5 gal   4 cups =1 quart 1 gill = .5 cup   4 tbsp = 1/4 cup

1 pinch = 1/8 tsp   1 pound = 454 grams 62 miles = 100 kilometers

1 ton =2000 pounds   metric ton = 1960 pounds

PLANE AND SOLID GEOMETRY ESSENTIALS

Appendix 5
Suggestions for further reading

Teacher: The author has used these books many more in various ways, one that was very impressive was to bring in books that the thought was suitable and enjoyable for the students to read the day before vacation. Let each student select a book, read it for a few minutes and if they didn't like then put it back and select another one. Many checked the selected book out and read it over vacation. But you have to know your students.

Abbott, Edwin A.
FLATLAND A ROMANCE OF MANY DIMENSIONS
Princeton University Press

Banks, Robert B.
SLICING PIZZAS, RACING TURTLES, AND FURTHER ADVANTURES IN APPLIED MATHEMATICS
Princeton University Press

Beckmann, P.
HISTORY OF PI
Golen Press

Bell, E. T.
MEN OF MATHEMATICS
Simon & Schuster

Byrkit, D.
TAXICAB GEOMETRY
MATHEMATICS TEACHER, May 1971, Pages 418-422

Cajori, Florian
HISTORY OF ELEMENTARY MATHEMATICS
The Macmillan Company

Davis, J.J.
BIBLICAL NUMEROLOGY
Baker Book House (PI value is stated in the Bible, erroneously, I Kings 7:23)

Davis, P. and Hersh, R.
THE MATHEMATICAL EXPERIENCE

Houghton Mifflin
DESCARTES DREAM
Harcourt Brace Javanovich

Devlin, K
Mathematics-the new golden age
Columbia University Press

Dudley, Underwood
NUNEROLOGY or, What Pythagoras Wrought
Mathematical Association of America
MATHEMATICAL CRANKS
Mathematical Association of America

Escher, M.C.
The Graphic Works
Printed in Germany

Fadiman, Clifton
THE MATHEMATICAL MAGPIE
(Mobius Strip- "Paul Bunyan vs. The Conveyor Belt)"
Simon and Schuster

Fawcett, Harold (For the teacher)
NATURE OF PROOF
13th Yearbook of NCTM

Florman, S. C.
ENGINEERING AND THE LIBERAL ARTS
McGraw-Hill Company
(A guide to History, Literature, Philosophy, Art, Science, and Music)

Gardner, M
MATHEMATICAL CARNIVAL
Alfred A. Knopf
Jacobs, W.R.
MATHEMATICS A HUMAN ENDEAVOR
W.H. FREEMAN & CO.

Jackson, Tom: Editor
Mathematics – An Illustrated History of Numbers
Shelter Harber Press

Knopf, Alfred A.
MATHEMATICAL CIRCUS
Vintage Books Division of Random House

Gazale, Midhat
NUMBER: From Ahmes to Cantor
Princeton University Press

Gordon, Sheldon and Florence, Editors
STATISTICS FOR THE TWENTY-FIRST CENTURY
Mathematical Association of America, 1992

Huff, Darrell
HOW TO LIE WITH STATISTICS
Norton Co.

Kenny
"Hemholtz And The Nature Of Geometric Axioms"
Mathematics Teacher, Vol. 50, Feb. 1957

Klein H. A.
THE WORLD OF MEASUREMENTS
Simon and Schuster

Kline, M.
MATHEMATICAL THOUGHT FORM ANCIENT TO MODERN TIMES
Oxford University Press

Lieber, L.
MITS, WITS, AND LOGIC
Institute Press, New York, 1954
THE EDUCATION OF T. C. MITS
W. W. Norton & Co., 1954

Loomis, E.
THE PYTHAGOREAN PROPOSITION.
NCTM publication
Comment: (Which former President of the U.S. is credited with a proof?)

Nolan, Deborah, Editor
WOMEN IN MATHEMTICS: SCALING THE HEIGHTS
Mathematical Association of America

Northrop, E. P.
RIDDLES IN MATHEMATICS (A Book of Paradoxes)
D. Van Nostrand Company

Packel, Edward
THE MATHAMATICS OF GAMES AND GAMBLING
Mathematical Association of America

Paulos, J.
I THINK, THEREFORE I LAUGH
Vintage Books
Division of Random House

Peterson, I.
THE MATHEMATICAL TOURIST
W. H. Freeman and Company

Poe, Edgar Allen
THE GOLD BUG (A Mystery involving mathematical reasoning)

Polya, G. (For the teacher)
MATHEMATICAL DISCOVERY: Vol. 2 (Chapter 14: The art of teaching mathematics.)
John Wiley & Sons

Postman, N.
TECHNOPOLY
Alfred A. Knopf

Reid, Constance
A LONG WAY FROM EUCLID
Thomas Y. Crowell Co.

Reeve, W. D. (For the teacher)
THE TEACHING OF GEOMETRY
5th Yearbook NCTM

Stevenson, R. L.
TREASURE ISLAND (chapter 31)
(Locus problem-location of the treasure.)

Weber, R.
A RANDOM WALK IN SCIENCE
Crane, Russak & Co. Inc.
"Life on Earth.(by a Martian")
(Fascinating little story (p. 124) with a surprise ending.

Videos or films
DONALD DUCK IN MATHMAGIC LAND
Disney

An interesting critical thinking test.
Critical Thinking Test, Level X
R. Ennis and J. Millman
(Very interesting and a different type of test based on a space travel theme.)
(The author has given this test to several hundred high school and college students on a pre/post test situation and to my surprise the average group gained the most.)
Available at:
Foundation for Critical Thinking 1-800-833-3645 or 1-800-458-4849
www.criticalthinking.org

Web sites

www.//history.mcs.st www.MAA.org
www.archives.math.utk.edu/societies.html
www.nsf.gov/www.AMS.org/

http://Turnbull.mcs.st
www.history.mcs.st-andrews
Also use computer search for "Math History"

Appendix 6
Quotes

Geometry with too much rigor only produces rigor mortise.

F. Allen

That they (all citizens) might excel in public discussions on philosophic or scientific questions, they must be educated (rhetoric, philosophy, mathematics, and astronomy).

The Athenian Sophist School Curriculum (480 B.C.E.)

Mathematics is the gate and the key to all the sciences. He who is ignorant of it cannot know he things of this world.

Roger Bacon

Neglect of mathematics works injury to all knowledge.

Roger Bacon

Mathematics - the unshaken Foundation of the Sciences, and the plentiful Fountain of Advantage to human affairs.

Issac Barrow

Students of mathematics ... the first time something new is studied they seem hopelessly confused...Then, upon returning (to the concept) after a rest, ...everything has fallen into place.

E. T. Bell
MEN OF MATHEMATICS

Hipparchus of Nicaea, (180-125 B.C.E.) compiled the first trigonometric table.

Boyer, C.B.
A HISTORY OF MATHEMATICS

I think therefore I am

Rene Descartes

Consciously Mathematics has been a human activity for thousands of years. To some small extent, everybody is a mathematician and does Mathematics

Phillip Davis & Reuben Hersh

THE MATHEMATICAL EXPERIENCE

Mathematics is about anything as long as it is a subject that exhibits the pattern of assumption-deduction-conclusion.

P. Davis and R. Hersh
The Mathematical Experience

The heart of the mathematical experience is, of course, mathematics itself.

P. Davis and R. Hersh
The Mathematical Experience

The theory of probability entered mathematics through gambling.

P. Davis and R. Hersh
The Mathematical Experience

It is easier to square the circle then get around a mathematician.

A. DE Morgan

The important thing is to not stop questioning.

Albert Einstein

Let no person ignorant of Geometry exit here.

J. Elander

It is not how much you cover, but how much you uncover.

H, Fawcett

There cannot be a language (mathematics) more universal... and more worthy to express the invariable relations of the natural things.

Joseph Fourier

In short, the house plays the percentages, while the player relies on luck.

H. Gross & F. Miller
MATHEMATICS--A CHRONICLE OF HUMAN
ENDEAVOR

In truth, all of life in one way or another is concerned with the study of probability.

H. Gross & F. Miller
MATHEMATICS - A Chronicle of Human
Endeavor

Mathematics through the power of computers pervades almost every aspect of our lives...

David L. Goines

God ever arithmetizes.

C.G. Jacobi

The Great Architect of the Universe now begins to appear as a pure mathematician.

J.H.Jeans
The Mysterious Universe

Mathematics through the power of computers pervades almost every aspect of our lives...

David L. Goines

Descartes...the essence of plane analytic geometry lies in the matching of ordered pairs of real numbers with points of a .plane.

Edna E. Kramer
THE NATURE AND GROWTH OF MODERN MATHEMATICS

A mathematician, like everyone else, lives in the real world. But the objects with which he works do not. They live in that other place--the mathematical world. Something else lives here also. It is called TRUTH.

Jerry P. King
THE ART OF MATHEMATICS

You cannot fake in mathematics, no one can be fooled. You can either prove(solve)... or you cannot.

Jerry P. King
THE ART OF MATHEMATICS

Attributing teaching and learning failure to something called "math anxiety" serves no purpose except to provide a built-in excuse for inadequate performance on both sides.

Jerry P. King
THE ART OF MATHEMATICS

GOD gave us the integers (whole numbers) and all the rest is the work of man.

L. Kronecker

TO MEASURE IS TO KNOW

Johann Kepler

YOUNG PEOPLE WHO HAVE ACQUIRED THE ABILITY TO ANALYZE PROBLEMS, GATHER INFORMATION, PUT THE PIECES TOGETHER TO FORM TENTATIVE SOLUTIONS WILL ALWAYS BE IN DEMAND.

J. G. Maisonrouge
Board Chairman
BM World Trade Corp.

Analytic Geometry...constitutes the greatest single step ever made in the progress of the exact sciences.

John Stuart Mill

There is no royal road to Mathematics

Menaechmus
(to Alexander the Great)

The advance and the perfecting of mathematics are closely joined to the prosperity of a nation.

Napoleon

Mathematics is the science of making necessary conclusions.

B. Peirce

Let no man ignorant of Geometry enter here. God ever Geometrizes

Plato

The connection between the improvement of human condition and the happiness of the human race is Science. (The Queen of the sciences is MATHEMATICS.)

Postman, N
Technopoly-The Surrender of
Culture to Technology

Just as statistics has spawned a huge testing industry, it has done the same for the polling of "public opinion."

Postman, N
Technopoly-The Surrender of
Culture to Technology

Statistics makes possible new perceptions and realities by making visible large-scale patterns.

Postman. N

Number rules the universe.

The Pythagoreans

Pythagoras, the teacher, paid his student three oboli (a coin) for each lesson he attended and noticed that as the weeks passed the boy's initial reluctance to learn was transformed into enthusiasm for knowledge. To test his pupil Pythagoras pretended that he could no longer afford to pay the student and that the lessons would have to stop, at which point the boy offered to pay for his education.

Simon Singh
FERMAT'S ENIGMA

Mathematics consists of islands of knowledge in a sea of ignorance. Relationships between different subjects (even branches of mathematics) are creatively important in mathematics.

Simon Singh
FERMAT'S ENIGMA

Statistical thinking will one day be as necessary for efficient citizenship as the ability to read and write.

H. G. Wells

The following interesting statements have been heard at conventions or hear say from others.

Mathematics is like a mighty tree with number (counting numbers) for its roots. Arithmetic grows on numbers, algebra on arithmetic, geometry on arithmetic and algebra, analytic geometry on arithmetic, algebra, and geometry. Calculus builds on all four. It is a tree that grows in time, fertilized by the minds of mathematicians and the applied needs of society

To Think is to Know.

Many of the laws of the sciences are stated in the language of variation.

We learn the new in the light of the old.

Thinkers recognize when two variables are related, but it is mathematics that connect them numerically.

We learn the new in the light of the old.

Statements are true or false. Conclusions are valid or invalid.

The proof of the pie is in the eating.

Mathematics is not a spectator sport!

Understanding evolves from work, appreciation from applications. Many of the laws of the sciences are stated in the language of variation.

Appendix 7
Pictures

I took many of these pictures when traveling, but unfortunately I did not record when or where. A few pictures involving interesting geometric designs are below.. Are there any in your community? Suggest some students may wish to start their own collection and show the class!

Simple church design

An older church design with bell tower

Modern Church Design

I hope some students took some very interesting pictures.
Jim Elander

Printed in the United States
by Baker & Taylor Publisher Services